T0135573

Delay Robustness in Cooperative Control

Von der Fakultät Konstruktions-, Produktions-, und Fahrzeugtechnik
der Universität Stuttgart zur Erlangung der Würde eines
Doktors der Ingenieurwissenschaften (Dr.–Ing.) genehmigte Abhandlung

Vorgelegt von

Ulrich Münz

aus Stuttgart-Bad Cannstatt

Hauptberichter: Prof. Dr.-Ing. Frank Allgöwer
Mitberichter: Dr. Antonis Papachristodoulou
Dr. Silviu-Iulian Niculescu, DR at CNRS (HDR)

Tag der mündlichen Prüfung: 1. Juni 2010

Institut für Systemtheorie und Regelungstechnik
Universität Stuttgart
2010

Bibliografische Information der Deutschen Nationalbibliothek

Die Deutsche Nationalbibliothek verzeichnet diese Publikation in der Deutschen Nationalbibliografie; detaillierte bibliografische Daten sind im Internet über http://dnb.d-nb.de abrufbar.

ISBN 978-3-8325-2591-0

Logos Verlag Berlin GmbH
Comeniushof, Gubener Str. 47,
10243 Berlin
Tel.: +49 (0)30 42 85 10 90
Fax: +49 (0)30 42 85 10 92
INTERNET: http://www.logos-verlag.de

Ἐπεὶ δὲ τὸ ἔκ τινος σύνθετον οὕτως ὥστε ἓν εἶναι τὸ πᾶν, [ἂν] μὴ ὡς σωρὸς ἀλλ' ὡς ἡ συλλαβή – ἡ δὲ συλλαβὴ οὐκ ἔστι τὰ στοιχεῖα, οὐδὲ τῷ ταὐτὸ τὸ καὶ (οὐδ' ἡ σὰρξ πῦρ καὶ γῆ) διαλυθέντων γὰρ τὰ μὲν οὐκέτι ἔστιν, [15] οἷον ἡ σὰρξ καὶ ἡ συλλαβή, τὰ δὲ στοιχεῖα ἔστι, καὶ τὸ πῦρ καὶ ἡ γῆ'·

Since that which is compounded out of something so that the whole is one, not like a heap but like a syllable — now the syllable is not its elements, ba is not the same as b and a, nor is flesh fire and earth (for when these are separated the wholes, i.e. the flesh and the syllable, no longer exist, but the elements of the syllable exist, and so do fire and earth).

Das was aus Bestandteilen so zusammengesetzt ist, daß es ein einheitliches Ganzes bildet, nicht nach Art eines Haufens, sondern wie eine Silbe, das ist offenbar mehr als bloß die Summe seiner Bestandteile. Eine Silbe ist nicht die Summe ihrer Laute; ba ist nicht dasselbe wie b plus a, und Fleisch ist nicht dasselbe wie Feuer plus Erde. Denn zerlegt man sie, so ist das eine, das Fleisch und die Silbe, nicht mehr vorhanden, aber wohl das andere, die Laute, oder Feuer und Erde.

Aristoteles (384 - 322 BC): Metaphysica VII 17, 1041 b

Acknowledgements

This thesis is the result of my five year dive into the academic world. This period of my life has been very intense, exciting, fruitful, and enriching. After all, I would like to thank several people who enabled and strongly supported my progress in this time.

My first thanks go to Prof. Dr.-Ing. Frank Allgöwer for supervising my thesis. He offered me a position as research and teaching assistant at the Institute for Systems Theory and Automatic Control with a much stronger emphasis on research than on teaching. Therefore, I was able to keep my mind focused on research. He opened several doors to the academic world for me, such as a summer school on time-delay systems. Moreover, he supported my trips to many international conferences and workshops in order to present my work, to see the most recent results of other researchers, and to get in contact with other scientist in my field. Moreover, he created a highly productive atmosphere at the institute with numerous national and international visitors and highly talented young researchers. On the other hand, he gave me the opportunity to study further on university didactics and improve my skills in classroom teaching and e-learning.

Second, I want to thank Dr. Antonis Papachristodoulou. He introduced me to the most exciting and challenging problems in networked systems. He pointed to those methodologies that could help to solve these problems. And he always encouraged me to continue on my way, even in the most tricky situations. Finally, I very much enjoyed our professional and personal friendship.

I would like to thank Dr. Silviu-Iulian Niculescu for his constructive comments on how to improve my thesis and for joining the committee of my PhD defense. I also thank him and Prof. Dr. ir. Wim Michiels for a very intensive and instructive one-week-course on time-delay systems in 2006. Both of them accompanied and supported my steps into the field of time-delay systems.

Moreover, I would like to thank Prof. Dr. Pedro Zufiria. He stimulated my interest in system theoretic problems; and he was the first one to show me the pleasure of solving such problems during our work on my master thesis.

My time at the institute would have been by far less relaxing, enjoyable, and productive without all my colleagues. They created an open atmosphere full of most interesting discussions on and off the job. I owe them many thanks for all I learned from them on many different control topics. Especially, I want to thank my office mates Jørgen Johnsen, Jan Hasenauer, Martin Löhning, and Peter Wieland, my collaborators on different research topics Tobias Schweickhardt, Peter Schumm, Christian Ebenbauer, Rainer Blind, Marcus Reble, Jochen Rieber, Christoph Maier, Gerd Schmidt, Christoph Böhm, and Simone Schuler, and the proofreaders of this thesis Simone Schuler, Matthias Bürger, Peter Wieland, Rainer Blind, Marcus Reble, Gerd Schmidt, and Steffen Waldherr. Moreover, I would like to thank my students over all the years who supported my research and teaching assistance: Angela Schöllig, Marcus Reble, Andreas Wiesebrock, Yi Liu, Thomas Haag, Ekaterini Kourti, Rainer Blind, Maxime Carré, Jing Jin, Sandro Bücheler, Johannes Eck, Jingbo Wu, and Gregor Goebel.

I would like to thank the Priority Programme 1305 *Control Theory of Digitally Networked Dynamical Systems* of the German Research Foundation (DFG) for their support and many interesting and fruitful discussions with fellow PhD students. I also would like to thank *The MathWorks Foundation in Science and Engineering at the University of Stuttgart* for their support of my research, in particular Bernd Kanamüller for interesting discussions on my work.

Last but not least, I would like to thank my family for their patience and support. Especially, my wife Kledy and my kids Jasmin and Leonhard had to cut back their own interests and I'm deeply indebted to them.

Contents

List of Figures

Symbols and Acronyms

Symbols

$\mathbb{R}$	set of reals
$\mathbb{R}^+(\mathbb{R}^-)$	set of positive (negative) reals
$\mathbb{R}_0^+(\mathbb{R}_0^-)$	set of non-negative (non-positive) reals
$\mathbb{C}^+(\mathbb{C}^-)$	set of complex numbers with positive (negative) real part
$\mathbb{C}_0^+(\mathbb{C}_0^-)$	set of complex numbers with non-negative (non-positive) real part
$\mathbb{C}_0^{+j}$	set of complex numbers with non-negative imaginary part
$\mathcal{C}_n([a,b],\mathbb{R}^n)$	Banach space of continuous functions mapping $[a,b]$ to $\mathbb{R}^n$, see Appendix A
$\mathcal{C}_n$	Banach space of continuous functions mapping $[-\mathcal{T},0]$ to $\mathbb{R}^n$
$\mathrm{Re}(z)$	real part of $z \in \mathbb{C}$
$\mathrm{Im}(z)$	imaginary part of $z \in \mathbb{C}$
v^*	complex conjugate of v
$\|\cdot\|$	Euclidean vector norm
$\lambda(\mathcal{M})$	eigenvalue of matrix $\mathcal{M}$
$\sigma(\mathcal{M})$	spectrum of matrix $\mathcal{M}$
$\mathcal{F}(\mathcal{M})$	field of values of matrix $\mathcal{M}$, see proof of Lemma 3.5 on page 38
$\det(\mathcal{M})$	determinant of matrix $\mathcal{M}$
$\mathrm{diag}(v_i)$	diagonal matrix with elements $v_i \in \mathbb{C}$
$y_{i[k]}(t)$	k-th element of $y_i(t)$
$y_i^{(\iota)}(t)$	ι-th derivative of $y_i(t)$
$\mathcal{L}(y(t))$	Laplace transform of $y(t)$
$\mathcal{L}^{-1}(Y(s))$	inverse Laplace transform of $Y(s)$
$\mathrm{Co}\{\cdot\}$	convex hull of a set
N	number of agents in the MAS
$\mathcal{N}$	index set of N agents
$\mathcal{N}_i$	index sets of all neighbours of agent i
$x_i(t)$	state of agent i
$u_i(t)$	input of agent i
$U_i(s)$	Laplace transform of $u_i(t)$
$y_i(t)$	output of agent i
$\overline{y}(t)$	average output of all agents
$Y_i(s)$	Laplace transform of $y_i(t)$
Σ_i	dynamics of agent i
$H_i = \frac{\nu_i(s)}{\delta_i(s)}$	transfer function of agent i

$k_{ij}(\cdot)$	coupling function between agent j and i
$\overline{k_{ij}}(\cdot)$	gain of coupling function $k_{ij}(z) = \overline{k_{ij}}(\\|z\\|)\frac{z}{\\|z\\|}$
$\Gamma_r(s)$	delay-interconnection matrix, see Equation (3.5) on page 23
$G_r(s)$	network return ratio, see Equation (3.21) on page 34
$\Omega_r(\omega\mathcal{T})$	convex sets containing $\sigma(2I - \Gamma_r(j\omega))$ for $\tau_{ij}, T_{ij} \leq \mathcal{T}$, see Equu tion (3.21) on page 34
$\Delta_0(s)$	characteristic polynomial of undelayed MAS
$\Delta(s)$	characteristic quasi-polynomial of MAS with delays
$\varphi \in \mathcal{C}_{\sum n_i}$	initial condition, $\mathcal{C}_{\sum n_i} = \mathcal{C}([-\mathcal{T}, 0], \mathbb{R}^{\sum n_i})$
$\mathcal{G} = (\mathcal{V}, \mathcal{E})$	graph
$\mathcal{V} = \{v_i\}$	vertex set
$\mathcal{E} \subseteq \mathcal{V} \times \mathcal{V}$	edge set
$\mathfrak{G} = \{\mathcal{G}_p\}$	finite sets of graphs $\mathcal{G}_p$
$\mathcal{G}([t_1, t_2])$	union graph over the interval $[t_1, t_2]$
h_{DW}	dwell-time
$A = [a_{ij}]$	adjacency matrix
$A_\tau(s)$	adjacency matrix including delays, see Equation (3.4) on page 23
$D = \mathrm{diag}(d_i)$	degree matrix
$D_\tau(s)$	degree matrix including identical self-delays, see Equation (3.4) o page 23
$D_T(s)$	degree matrix including different self-delays, see Equation (3.4) o page 23
$L = D - A$	Laplacian matrix
$\overline{L} = I - D^{-1}A$	normalized Laplacian matrix
$L_\tau(s)$	Laplacian matrix including identical self-delays, see Equation (3.5) o page 23
$L_{\tau T}(s)$	Laplacian matrix including different self-delays, see Equation (3.5) o page 23
λ_2	algebraic connectivity, second smallest eigenvalue of $\overline{L} = I - D^{-1}A$
$\overline{\lambda}_2$	lower bound on the second smallest eigenvalue of $\overline{L} = I - D^{-1}A$
$\mathbb{T}_{ij} : \mathcal{C}_m \to \mathbb{R}^m$	delay operator for delay from agent j to agent i, see Section 2.6
$\overline{\mathbb{T}}_{ij} : \mathcal{C}_m \to \mathbb{R}^m$	delay operator for self-delay of agent i, see Section 2.6
τ_{ij}	constant delay in the channel from j to i
T_{ij}	self-delay of agent i with respect to agent j
$\tau_{ij}(t)$	time-varying delay in the channel from j to i
$\mathcal{T}_{ij}$	upper bound of time-varying delay $\tau_{ij}(t)$
$\phi_{ij}(\eta)$	kernel of the distributed delay from agent j to agent i, see Equu tion (2.16) on page 18
$\overline{\tau}_{ij}$	average delay of distributed delay
$\mathcal{T}$	upper bound of all delays, delay range
$y_{i,t} \in \mathcal{C}_m$	segment of trajectory $y_i(t), t \in [t - \mathcal{T}, t]$
$y_{i,t}(\eta)$	$= y_i(t + \eta), \eta \in [-\mathcal{T}, 0]$, element of segment $y_{i,t}$
Υ	hyper-rectangle for contraction argument, see Section 4.2

$\partial\Upsilon$	boundary of set Υ
$M_i(y_i)$	inertia matrix
$C_i(y_i, \dot{y}_i)\dot{y}_i$	centrifugal and Coriolis forces
$S_i(x_i)$	storage function of agent i

Acronyms

CDMA	code-division multiple access
LTI	linear, time-invariant
MAS	multi-agent system
OVM	optimal velocity model
PSCN	packet-switched communication network
RD1, RD2	relative degree one, relative degree two
RFDE	retarded functional differential equation
TDS	time-delay system

Abstract

The robustness of various cooperative control schemes on large scale networked systems with respect to heterogeneous communication and coupling delays is investigated. The presented results provide delay-dependent and delay-independent conditions that guarantee consensus, rendezvous, flocking, and synchronization in different classes of multi-agent systems (MAS). All conditions are scalable to arbitrarily large multi-agent systems with non-identical agent dynamics. In particular, conditions for linear agents, for nonlinear agents with relative degree one, and for a class of nonlinear agents with relative degree two are presented. The interconnection topology between the agents is in most cases represented by an undirected graph. The results for nonlinear agents with relative degree one hold also for the more general case of directed graphs with switching topologies. Different delay configurations are investigated and compared. These configurations represent different ways how the delays affect the coupling between the agents. The presented robustness analysis considers constant, time-varying, and distributed delays in order to take different sources of delays into account. The results are applied to several typical applications and simulations illustrate the findings.

The main contributions of this thesis include: (i) Consensus and rendezvous in single integrator MAS are robust to arbitrarily large delays even on switching topologies. However, the convergence rate of this MAS is delay-dependent and scalable convergence rate conditions are presented. (ii) Consensus and rendezvous in relative degree two MAS are robust to sufficiently small delays. Local, scalable conditions are derived for these MAS that guarantee consensus and rendezvous for bounded delays. (iii) Finally, the derived delay robustness analysis for general linear MAS allows for the first time to compare different delay configurations in a unifying framework.

Deutsche Kurzfassung (German Abstract)

In dieser Dissertation wird die Robustheit von verschiedenen kooperierenden Regelungen bezüglich heterogener Kommunikations- und Kopplungsverzögerungen untersucht. Es werden skalierbare Analysemethoden für Konsens, Rendezvous, Herdenverhalten (engl. flocking) und Synchronisation für verschiedene Klassen von linearen und nichtlinearen Agentensystemen (engl. multi-agent systems) vorgestellt. Die Ergebnisse werden auf verschiedene typische Anwendungen übertragen und durch Simulationen veranschaulicht.

Motivation

Bei vielen Phänomenen in der Natur spielt die Kooperation zwischen Individuen einer Gruppe eine wesentliche Rolle. Beispielhaft seien hier das Verhalten von Fisch- und Vogelschwärmen, die Nahrungsbeschaffung bei Ameisenvölkern sowie die Synchronisation der Nervenzellen, die den Herzschlag vorgeben, genannt. Eine gute Einführung in dieses Thema mit zahlreichen weiteren Beispielen gibt Strogatz (2004) (englische Originalfassung Strogatz (2003)). Es ist eine wesentliche Eigenschaft dieser Kooperation, dass das Verhalten der Gruppe nicht von einzelnen Individuen vorgegeben wird. Vielmehr ergibt sich das Verhalten aufgrund der lokalen Interaktion zwischen jedem Mitglied der Gruppe und dessen Nachbarn. So orientiert sich jeder Fisch in einem Schwarm an den Bewegungen der anderen Fische in seiner Umgebung. Aber er weiß nicht, wohin der Schwarm als ganzes schwimmt. Dennoch bleiben die Fische im Schwarm und bewegen sich gemeinsam in eine bestimmte Richtung (Couzin et al., 2005; Pöppe, 2005; Reynolds, 1987).

Zahlreiche technische Systeme bestehen ebenfalls aus vielen kooperierenden dynamischen Teilsystemen. Beispielhaft seien hier drei Anwendungen genannt: Kommunikationsnetze wie das Internet setzen sich aus vielen Routern zusammen, die Information von Millionen von Quellen zu ebenso vielen Anwendern übertragen und dabei die Kapazität der einzelnen Übertragungskanäle berücksichtigen (Srikant, 2004). Verkehrsnetze bestehen aus vielen Zügen, Fahrzeugen oder Flugzeugen mit dem gemeinsamen Ziel, Passagiere und Waren von einem Ort zum anderen zu bringen (Helbing, 2001). In Energieversorgungsnetzen kooperieren die Generatoren um im gesamten Netz eine konstante Spannung und Frequenz unabhängig von der Zahl der Verbraucher zu gewährleisten (Pavella and Murthy, 1994). Weitere Anwendungen werden in Murray (2002) vorgestellt.

Diese kooperierenden dynamischen Systeme werden in der englischsprachigen Literatur *multi-agent systems (MAS)* genannt. Im deutschen Sprachgebrauch hat sich der Begriff *Agentensystem* in Anlehnung an kooperierende Systeme in der Automatisierungstechnik etabliert. In der Automatisierungstechnik bezeichnen Agentensysteme Gruppen von Agenten, die interagieren, um gemeinsame Ziele zu erreichen, siehe VDI-Richtlinie

„Agentensysteme in der Automatisierungstechnik" ((VDI), 2009). Diese Definition entspricht im wesentlichen dem Gebrauch des Begriffs multi-agent system in der englischsprachigen Literatur. Einen Überblick über dieses schnell wachsende Gebiet der Regelungstechnik geben die Sonderausgaben verschiedener Zeitschriften (Antsaklis and Baillieul, 2007; Roy et al., 2007; Hu et al., 2008; Shima and Pagilla, 2007) sowie die Bücher (Ren and Beard, 2008; Qu, 2009; Shamma, 2007; Shima and Rasmussen, 2009; Bullo et al., 2009).

Agentensysteme in der Regelungstechnik haben zwei charakteristische Eigenschaften:

- Sie bestehen aus einer Vielzahl von dynamischen Teilsystemen, deren Anzahl sich mit der Zeit ändern kann, und
- die genaue Verknüpfungstopologie zwischen den Agenten ist normalerweise unbekannt und manchmal ebenfalls zeitvariant.

Diese Eigenschaften machen in den meisten Fällen eine zentrale Regelung des Agentensystems unmöglich. Daher versuchen Ingenieure von der Natur zu lernen, wie kooperierende Regelungen für Agentensysteme entworfen werden können. Diese Regelungen sollen ein globales Verhalten der Gruppe durch lokale Interaktion zwischen den Agenten garantieren. Man spricht in diesem Zusammenhang oft von *Schwarmintelligenz* (Bischoff, 2009; Miller, 2007b) (englische Originalversion Miller (2007a)).

Die Interaktion zwischen den Agenten erfordert einen Informationenaustausch über ein Netzwerk. Dieser Informationsaustausch kann ein inhärenter Teil des Agentensystems sein, wie z.B. die Phasenkopplung zwischen Generatoren in einem Energieversorgungsnetz. Bei anderen Anwendungen tauschen die Agenten Informationen über ein digitales Kommunikationsnetzwerk aus, wie z.B. bei der aktiven Warteschlangenregelung im Internet (engl. active queue management (AQM)). Eine dritte Form des Informationsaustauschs sind Messungen, wie z.B. Abstandsmessungen bei Fahrzeugkolonnen. In den meisten Fällen wird die Information beim Datenaustausch verzögert, sei es durch die endliche Ausbreitungsgeschwindigkeit in Energieversorgungsnetzen, durch Warteschlangen in digitalen Netzen oder durch Messzeiten bei Abstandsmessungen. Daher sind diese Verzögerungen wesentlicher Bestandteil vieler Agentensysteme. Auch wenn diese Verzögerungen sehr klein sind, können sie ein kooperatives Verhalten verhindern. Ein Beispiel hierfür wird in Kapitel 5.4 vorgestellt. Daher sind Methoden für die Robustheitsanalyse von kooperierenden Systemen bezüglich dieser Totzeiten[1] sehr wichtig. Eine ausführliche Übersicht über den Forschungsstand in diesem Bereich befindet sich in der englischsprachigen Einleitung in Kapitel 1.

In dieser Arbeit werden die elementarsten kooperative Verhaltensweisen eines Agentensystems auf ihre Robustheit bezüglich Totzeiten untersucht. Diese sind

- *Konsens*, z.B. wenn Mikroprozessoren mit lokalem Takt sich auf einen einheitlichen Takt einigen,

[1] In der Nachrichtentechnik bezeichnet man Zugriffs- und Übertragungszeiten als Verzögerungen. Diese Effekte werden in der Systemtheorie Totzeiten genannt. In dieser Kurzfassung werden Fachbegriffe beider Fachbereiche verwendet. Daher werden die Begriffe Verzögerung und Totzeit synonym verwendet.

- *Rendezvous*, z.B. wenn Roboter sich an einem Punkt treffen oder eine Formation einnehmen,
- *Herdenverhalten* (engl. flocking), z.B. wenn Fahrzeuge in einer Kolonne mit der gleichen Geschwindigkeit und einem bestimmten Abstand zueinander fahren und
- *Synchronisation*, z.B. die Phasensynchronisation der Generatoren in Energieversorgungsnetzen.

Eine formale Definition dieser Eigenschaften wird in Kapitel 2 gegeben. Komplexere kooperierende Verhaltensweisen können meist aus einer dieser vier elementaren Verhaltensweisen abgeleitet werden.

Um Aussagen über beliebig große Agentensysteme treffen zu können, darf die Komplexität der Robustheitsanalyse nur leicht oder überhaupt nicht mit der Zahl der Agenten, der Zahl der Verbindungen zwischen den Agenten und der Zahl der Totzeiten zunehmen. Diese Eigenschaft wird *Skalierbarkeit* genannt, siehe z.B. Bondi (2000). Darüber hinaus ist es wünschenswert, dass die Robustheitsbedingungen lokale Bedingungen sind, d.h. sie setzen die Dynamik jedes Agenten mit der Dynamik seiner benachbarten Agenten und den Verbindungen zu diesen Agenten in Beziehung. Gleichzeitig sind diese Bedingungen unabhängig von der Dynamik anderer Agenten oder von anderen Verbindungen im Netzwerk. Somit stellen lokale Bedingungen die Idealform skalierbarer Bedingungen dar. Sie erlauben einen lokalen Entwurf der Regler anstelle eines globalen Entwurfs. Dies ist insbesondere dann von Vorteil, wenn Teilsysteme zu dem Agentensystem hinzustoßen oder dieses verlassen. Wenn zum Beispiel ein neuer Generator mit einem Energieversorgungsnetz verbunden wird, ist es nicht möglich, die Regler aller Generatoren im Energieversorgungsnetz neu zu entwerfen. Stattdessen sollte nur der Regler des neuen Generators und gegebenenfalls die Regler seiner Nachbarn angepasst werden.

Gliederung und Forschungsbeiträge dieser Dissertation

In diese Dissertation werden verschiedene Methoden zur skalierbaren Robustheitsanalyse von kooperierenden Reglern bezüglich Totzeiten entwickelt. Die Agentensysteme bestehen aus Agenten mit zeitkontinuierlichen, linearen oder nichtlinearen Dynamiken. Die Verbindungen zwischen den Agenten wird mittels eines Graphen beschrieben, der in den meisten Fällen ungerichtet ist. Die Ergebnisse in Kapitel 4 gelten allerdings auch für den allgemeineren Fall von gerichteten und schaltenden Graphen. Darüber hinaus werden verschiedene Konfigurationen untersucht und verglichen, welche sich darin unterscheiden, wie die Verzögerungen die Interaktion zwischen den Agenten beeinflussen. Dabei werden konstante, zeitvariante und verteilte Totzeiten betrachtet, um unterschiedliche Quellen für die Totzeiten zu berücksichtigen. In der gesamten Arbeit werden heterogene Totzeiten angenommen, d.h. die Totzeiten können in jeder Verbindung zwischen zwei Systemen unterschiedlich sein.

Weite Teile dieser Dissertation befassen sich mit Konsens und Rendezvous in Agentensystemen. Dies ist teilweise der Tatsache geschuldet, dass viele Agentensysteme mit Totzeiten kein Herdenverhalten erzielen können, was ebenfalls in dieser Arbeit näher untersucht wird. In Anwendungsbeispielen wird gezeigt, wie die vorgestellten Methoden dennoch verwendet werden können, um die Synchronisation von Kuramoto-Oszillatoren (Kuramoto, 1984) oder das Folgeverhalten von Fahrzeugen (Helbing, 2001) zu untersuchen. Eine detaillierte Einführung in die untersuchten Agentendynamiken, Verknüpfungstopologien, Totzeitmodelle und kooperative Regelziele wird in Kapitel 2 gegeben.

Für dieses breite Spektrum an Agentendynamiken und Regelzielen werden drei unterschiedliche Methoden basierend auf dem verallgemeinerten Nyquistkriterium, einem Kontraktionsargument und Summen positiv definiter Lyapunov-Krasovskii Funktionale entwickelt. Jedes der nachfolgenden Kapitel dieser Arbeit ist einer dieser Methoden gewidmet. Tabelle 1.1 auf Seite 3 gibt einen Überblick über die betrachteten Agentensysteme in diesen Kapiteln. Die Forschungsbeiträge in den einzelnen Kapiteln sind im folgenden zusammengefasst:

Kapitel 3: Verallgemeinertes Nyquistkriterium Es wird eine skalierbare, mengenbasierte Methode für die Robustheitsanalyse von Konsensproblemen in linearen Agentensystemen entworfen. Die Agenten sind über Netzwerke verknüpft, die durch ungerichtete, konstante Graphen und konstante Totzeiten beschrieben werden. Drei unterschiedliche Konfigurationen, wie die Totzeiten die Verbindungen zwischen den Agenten beeinflussen, werden untersucht: Totzeiten, die nur die Informationen der Nachbaragenten betreffen, und Totzeiten, die sowohl die Nachbarn als auch die Informationen des Agenten selbst betreffen, wobei im zweiten Fall unterschieden wird, ob die Totzeiten der Nachbarn und des Agenten selbst identisch sind oder nicht. Mithilfe dieser mengenbasierten Methode werden Robustheitsbedingungen für verschiedene Klassen linearer MAS entwickelt, z.B. für Agentensysteme bestehend aus Einfachintegratoren. Darüber hinaus wird eine skalierbare Bedingung für die Konvergenzrate von Agentensystemen bestehend aus Einfachintegratoren hergeleitet.

Kapitel 4: Kontraktionsargument Es wird gezeigt, dass nichtlineare Agenten mit relativem Grad eins und beliebig langen Verzögerungen ein Rendezvous erreichen, sofern die Kopplung zwischen den Agenten ein integrierendes Verhalten garantiert. Dabei werden Agenten betrachtet, die ihren eigenen Ausgang ohne Verzögerung bestimmen können. Unter den drei in Kapitel 3 untersuchten Konfigurationen ist dies die robusteste gegen Totzeiten. Das Rendezvous dieser Systeme wird mithilfe eines Kontraktionsarguments für ein allgemeines Netzwerk mit gerichteten, schaltenden Topologien und konstanten, zeitvarianten oder verteilten Totzeiten nachgewiesen. Ausgehend von dieser sehr allgemeinen Systemklasse werden Bedingungen für wichtige Unterklassen des betrachteten Agentensystems abgeleitet, wie Agenten mit Euler-Lagrange-Dynamik oder Einfachintegratoren. Schließlich wird sowohl theoretisch gezeigt als auch durch Simulationen veranschaulicht, dass Kuramoto-Oszillatoren unter gewissen Voraussetzungen auch dann synchronisieren, wenn die Kopplungen mit Übertragungsverzögerungen behaftet sind.

Kapitel 5: Summe positiv definiter Lyapunov-Krasovskii Funktionale Es werden lokale, skalierbare, totzeitabhängige Bedingungen an die Kopplungsfunktionen hergeleitet, so dass für eine große Klasse von nichtlinearen Agenten mit relativem Grad zwei Rendezvous garantiert werden kann. Dabei wird angenommen, dass der zugrundeliegende Graph konstant und ungerichtet ist. Die betrachtete Klasse von nichtlinearen Agenten mit relativem Grad zwei ist praxisrelevant, sie beinhaltet beispielsweise Euler-Lagrange Systeme. Ähnlich wie in Kapitel 4 werden konstante, zeitvariante und verteilte Totzeiten berücksichtigt, wobei der Agent seinen eigenen Ausgang ohne Totzeit bestimmen kann.

Die vorgestellten Bedingungen werden mithilfe von Summen positiv definiter Lyapunov-Krasovskii Funktionale hergeleitet. Diese Bedingungen garantieren darüber hinaus eine Robustheit gegenüber Modellunsicherheiten und hängen lediglich von lokalen Totzeiten ab, d.h. von einer oberen Schranke für die Totzeiten zu den Nachbarn jedes Agenten aber nicht von allen Totzeiten im gesamten Netzwerk. Daher sind diese Rendezvousbedingungen gut für große Netzwerke mit unterschiedlichen Agentendynamiken geeignet.

Ausgehend von diesem allgemeinen Resultat werden Bedingungen für spezielle Systemklassen abgeleitet, wie Fahrzeugfolgemodelle oder Agenten mit Euler-Lagrange-Dynamik. Darüber hinaus werden die Ergebnisse mit ähnlichen Bedingungen basierend auf dem verallgemeinerten Nyquistkriterium in Kapitel 3 verglichen. Schließlich werden die Ergebnisse anhand von Simulationen veranschaulicht. Diese zeigen, dass auch sehr kleine Verzögerungen ein Rendezvous von Agenten verhindern können. Dabei wird ebenfalls deutlich, wie die hier vorgestellten Methoden helfen können, kooperierende Regler so zu entwerfen, dass sie robust gegenüber Totzeiten sind.

Zusammenfassung

Diese Dissertation umfasst mehrere Methoden, mit denen die Robustheit von kooperierenden Reglern bezüglich Totzeiten untersucht werden kann. Diese Methoden decken ein breites Spektrum an unterschiedlichen Agentendynamiken, kooperativen Regelzielen, Netzwerktopologien, sowie Totzeitmodellen und -konfigurationen ab. Darüber hinaus sind die Ergebnisse skalierbar auf beliebig große Agentensysteme mit heterogenen Totzeiten. Eine ausführliche Diskussion der Ergebnisse dieser Dissertation und ein Ausblick auf zukünftige Forschungsthemen in diesem Bereich finden sich in Kapitel 6.

Chapter 1

Introduction

1.1 Motivation

Cooperative behavior in large groups of individuals can be found abundantly in nature. Well-known examples are schools of fish, flocks of birds, collective food-gathering in ant colonies, as well as synchronization of flashing fireflies and pacemaker cells, see Strogatz (2003) for a nice introduction with many examples. A fundamental property of this cooperation is that the group behavior is not dictated by one of the individuals. On the contrary, this behavior results implicitly from the local interaction between the individuals and their neighbours. For instance, every fish in a school knows where the other fish in its neighbourhood are heading, but it does not know the average heading of all fish in the school. Nonetheless, the fish in the school stay together and move as a group in a certain direction (Couzin et al., 2005; Reynolds, 1987).

Many engineering systems also consist of large groups of cooperating dynamic systems. They are called *multi-agent systems* (MAS) in the literature, see Olfati-Saber et al. (2007); Ren and Beard (2008) for recent overviews. Various applications are provided in Murray (2002); we mention here only three examples: Communication networks like the Internet are composed of many routers with the aim to transmit information from millions of sources to equally many users respecting the capacity of each link of the network (Srikant, 2004). Transport systems consist of many trains, cars, or airplanes with the common aim to bring people and goods from one point to another, see Helbing (2001) for an overview on car-following. In power networks, the power generators have to cooperate in order to provide a constant voltage and frequency irrespective of how many consumers are connected to the network (Pavella and Murthy, 1994). These applications show two main characteristics of MAS:

- the group consists of a large number of subsystems and their number may even be time-varying as new agents join or leave the group, and
- the interconnection topology between the agents is usually unknown and changing over time.

These properties often render a centralized control of the MAS very difficult. Therefore, engineers seek to learn from nature how to implement a *decentralized cooperative control* strategy that achieves global goals based on local couplings (Miller, 2007a).

These cooperative control strategies require that the agents exchange information with their neighbours over a network. This exchange of information can be an implicit property of the MAS, e.g. the inherent phase coupling of power plants in power grids. In other applications, the agents exchange information using digital communication networks in order to fulfill a cooperative control task, e.g. in Internet congestion control. Finally, there are applications where the information is exchanged using measurements, e.g. relative distance measurements in multi-vehicle applications. Many of these networks introduce some sort of delay when information is exchanged between the agents, either *propagation delays* as in power grids, or *access* and *queuing delays* in digital communication networks, or *measurement delays* as in the multi-vehicle example. These delays are an intrinsic property of many MAS. Even though these delays are often small, they might impede that the MAS fulfills the cooperative control task; an example is given in this thesis. Therefore, the robustness of cooperative control strategies with respect to delays is very important.

We investigate in this thesis the delay robustness of the most fundamental cooperative control tasks:

- *consensus*, e.g. micro controllers with local clocks agreeing on a common time,
- *rendezvous*, e.g. robots meeting at a point or achieving a certain formation,
- *flocking*, e.g. cars moving with the same velocity and with a certain distance between each other, and
- *synchronization*, e.g. phase synchronization of generators in power networks.

A formal definition of these cooperative control tasks is given in Chapter 2. More sophisticated tasks can be usually reduced to one of these problems.

In order to provide conditions for arbitrarily large networks, the complexity of delay robustness analysis should be independent of the number of agents, the number of interconnections, and the number of delays, or it may only increases very slowly with these parameters. This property is called *scalability*, e.g. Bondi (2000). Moreover, it is desirable that delay robustness conditions are *local conditions*, i.e. they relate the dynamics of every agent to its neighbouring agents and the interconnections to these neighbours. Yet, these conditions are independent of other agents or other interconnections in the network. Thus, local conditions are the ideal case of scalable conditions. They allow for a local design of the couplings, avoiding a global design of the whole MAS. This is particularly useful if some agents join or leave the network. For instance, if a new power plant is connected to a power grid, it is not viable to redesign the controllers of all power plants in the network.

1.2 Focus and Contributions of this Thesis

We present scalable methods for delay robustness analysis in cooperative control. We consider various cooperative control tasks for a wide spectrum of MAS with different

Table 1.1: Overview of different MAS models and cooperative control tasks in this thesis.

	agent dynamics	cooperative control task	self-delay	graph	delay models
Ch. 3	linear	consensus, flocking	no, identical, different	undirected, constant	constant
Ch. 4	nonlinear, RD1, integrating behavior	consensus, rendezvous, synchronization	no	directed, switching	constant, time-varying, distributed
Ch. 5	nonlinear, RD2, special structure	consensus, rendezvous, flocking	no	undirected, constant	constant, time-varying, distributed

kinds of delays. All results are scalable to large MAS. Moreover, we present several local conditions for delay robustness. The methods are applied to several practical examples and illustrated in simulations.

More precisely, we consider MAS with continuous-time, linear or nonlinear agent dynamics. The topology of the network between the agents is described in most cases by an undirected graph. The results in Chapter 4 hold also for the more general case of directed, switching graphs. Moreover, we take different delay configurations into account that reflect different ways how the delays affect the coupling between the agents. In addition, we consider constant, time-varying, and distributed delay models to cover different sources of delays, like access delays, propagation delays, and measurement delays. Throughout this thesis, we consider heterogeneous delays, i.e. the delays can be different for each interconnection link.

Large parts of this thesis concentrate on consensus and rendezvous in MAS. This is also due to the fact that many MAS with delays cannot achieve flocking and synchronization with the standard cooperative control laws, as will be shown in this thesis. In addition, we show exemplarily how the derived results can be used to investigate synchronization of Kuramoto oscillators and flocking in car-following models. A more detailed introduction to the agent dynamics, the interconnection topologies, the delay models, as well as the cooperative control tasks is given in Chapter 2.

In order to cope with such a broad range of MAS and cooperative control tasks, we develop three different methodological approaches based on the generalized Nyquist criterion, a contraction argument, and sums of Lyapunov-Krasovskii functionals. Each of the subsequent chapters is dedicated to one of these approaches. An overview of the considered MAS in these chapters is given in Table 1.1. The contributions of the individual chapters are summarized next:

Chapter 3: Generalized Nyquist Criterion We develop a set-valued method to investigate the delay robustness of consensus and flocking in linear MAS on undirected graphs with heterogeneous, constant delays. We consider three types of output feedback delays: delays affecting only the output of the agents' neighbours as well as delays affecting both the agents' own output and the output of their neighbours, distinguishing whether the self-delays and the neighbours' delays are identical or different. The applicability of the set-based method is demonstrated for several special cases like consensus in MAS with single integrator agent dynamics and consensus controller design for a class of linear MAS. Moreover, we derive scalable and robust convergence rate conditions for MAS with single integrator agent dynamics. The main contributions of this chapter are:

- a unifying framework for delay robustness analysis of linear MAS with different delay configurations,
- consensus conditions for single integrator MAS with different delay configurations, and
- convergence rate conditions for single integrator MAS with heterogeneous delays.

Chapter 4: Contraction Argument We show that MAS consisting of nonlinear agents with relative degree one (RD1) eventually reach output rendezvous for arbitrarily large delays if the coupling functions between the agents guarantee an integrating behavior. In this chapter, we concentrate on MAS without self-delay. Among the three delay configurations in Chapter 3, this is the most robust one to delays. Rendezvous is proven using a contraction argument. Thereby, we consider the most general network with constant, time-varying, or distributed delays and switching, directed topologies. From this general result, we derive rendezvous controller for MAS with nonlinear, input affine agents, for agents with Euler Lagrange dynamics, and for single integrator agent dynamics. The latter corresponds to the standard MAS for rendezvous and consensus in the literature, e.g. Olfati-Saber et al. (2007); Ren and Beard (2008). Finally, we show theoretically and by simulations that delay-coupled, identical Kuramoto oscillators synchronize. The main contribution of this chapter is:

- proof of delay-independent rendezvous in nonlinear MAS with integrating behavior under the same conditions as for undelayed MAS, even if the underlying graph is directed and switching.

Chapter 5: Sums of Lyapunov-Krasovskii functionals We provide local, scalable, delay-dependent conditions on the coupling functions such that MAS consisting of a class of nonlinear agents with relative degree two (RD2) achieve rendezvous on constant, undirected graphs. The class of nonlinear RD2 agents is relevant for rendezvous problems and includes, for example, systems with Euler Lagrange dynamics. As in Chapter 4, we concentrate on delay configurations without self-delay and take all three delay models into account: constant, time-varying, and distributed delay.

The conditions are derived using sums of Lyapunov-Krasovskii functionals. These conditions are robust to model uncertainties and only depend on local parameters, e.g. on upper bounds on the delays to the agent's neighbours but not on a global upper bound on the delays in the whole network. Thus, these local rendezvous conditions are scalable to large networks of non-identical agents.

From this general result, we derive specific conditions for Euler Lagrange MAS and car-following models. Moreover, we compare these conditions to similar conditions for MAS composed of linear agents with relative degree two derived in Chapter 3. Finally, simulation results illustrate that even very small delays may impede rendezvous. These simulations also show how rendezvous can be guaranteed in these cases using the methods presented in this thesis. The main contributions of this chapter are:

- local, delay-dependent rendezvous conditions for MAS consisting of a class of nonlinear agents with relative degree two and
- application of these results to Euler Lagrange MAS and a car-following model.

In summary, we present in this thesis a variety of methods and conditions for a scalable delay robustness analysis of different cooperative control tasks in a broad range of MAS.

1.3 Related Work

Cooperative behavior has been studied for the first time more than 300 years ago when Christiaan Huygens discovered synchronization of pendulum clocks (Huygens, 1673). However, the mechanisms behind this and other synchronization phenomena have only been revealed in the last 50 years, mainly using computer simulations and strong simplifications, see Strogatz (2003) for a nice review. One of the first publications on consensus dates back roughly 50 years, when Eisenberg and Gale (1959) investigated the consensus of individual probability distributions of bettors on horse races. Early consensus conditions for linear discrete-time MAS based on stochastic matrices have been published in DeGroot (1974). A nice overview of early works on cooperating systems is given in Vámos (1983) and one of the first PhD theses on this topic in the control community is Tsitsiklis (1984).

In the last five years, cooperative control has become one of the most active research areas in the control community. It is beyond the scope of this thesis to review all important publications in this area. The interested reader is referred to following special issues and books: Antsaklis and Baillieul (2007); Roy et al. (2007); Hu et al. (2008); Shima and Pagilla (2007); Ren and Beard (2008); Qu (2009); Shamma (2007); Shima and Rasmussen (2009); Bullo et al. (2009). Our literature review will be focused on MAS with delays.

Consensus in discrete-time single integrator MAS with linear coupling and delays has been studied extensively in the past. The main publications show that consensus is reached on uniformly quasi-strongly connected graphs[1], even if the agents communicate asynchronously, see Xiao and Wang (2008a,b); Cao et al. (2008); Fang and Antsaklis (2005); Almeida et al. (2009). In order to derive these results, the state space is extended by delayed states. Then, results on ergodic matrices are applied in order to prove consensus. Similar methods have been applied to discrete-time second order MAS in Liu and Passino (2006); Lin and Jia (2009). Alternative approaches use the generalized

[1] See Definition 4.2 on page 63 for more details on uniformly quasi-strongly connected graphs

Nyquist criterion and Gershgorin's circle theorem to proof consensus (Tian and Liu, 2008). Generally speaking, the main results on delay robustness analysis for discrete-time MAS heavily rely on the fact that these MAS are finite dimensional systems. Thus, these methods are not applicable to infinite dimensional systems like continuous-time MAS with delays. In Chapter 4, we generalize these results to continuous-time MAS with delays using a contraction argument.

In continuous time, delay robustness in cooperative control can be investigated in the frequency domain, e.g. using small gain arguments or the generalized Nyquist criterion, and in the time-domain, e.g. using Lyapunov-Krasovskii functionals or Lyapunov-Razumikhin functions. We first review different results in the frequency domain which are always restricted to linear MAS. In this case, frequency- and delay-dependent feedback matrices are defined depending on the topology and the heterogeneous delays. Then, convex sets are determined which contain the eigenvalues of the feedback matrices for arbitrary topologies and bounded delays. One possibility to compute such a set is Gershgorin's circle theorem, e.g. Horn and Johnson (1985). Consensus conditions based on Gershgorin's circle theorem typically apply a small gain condition and lead to delay-independent results because Gershgorin's circles contain the eigenvalues of the feedback matrices for arbitrary large delay values, e.g. Wang and Elia (2008); Lee and Spong (2006); Charalambous et al. (2008); Stefanovic and Pavel (2009a). In Liu and Tian (2007), the generalized Nyquist criterion is combined with Gershgorin's circles and consensus conditions are derived for single integrator MAS with arbitrary large communication delays but bounded computation delays. In Lestas and Vinnicombe (2007b,a, 2006), frequency-dependent sets are determined and set-valued stability and consensus conditions are derived similar to the results in Chapter 3. This approach uses similar methods as our results in Chapter 3, a detailed comparison is given in Section 3.4. The method proposed in Chapter 3 develops relatively simple frequency-dependent and delay-dependent convex sets that contain the eigenvalues of the feedback matrix. These sets are much more accurate than Gershgorin's circles. The resulting consensus conditions based on these sets also apply for a much more general class of linear MAS with delays and, hence, contain most previous results as special case. In particular, these conditions consider for the first time different delay configurations in a unifying framework and provide robust convergence rate conditions for MAS with heterogeneous delays.

In the time-domain, delay robustness of MAS has been studied using small-μ analysis (Yang et al., 2008), input-to-state stability (Fan and Arcak, 2006), or techniques for partial differential equations (Bliman and Ferrari-Trecate, 2008). In Kao et al. (2009); Jönsson and Kao (2009), parameterized convex sets are defined and it is assumed that the eigenvalues of the feedback matrix are contained in this convex set. The feedback matrix is defined in a similar way as in the frequency-domain approaches mentioned above. Then, stability conditions are derived based on integral quadratic constraints (IQC) using the parameters of this convex set. In order to determine the parameters of this convex sets, the authors propose Gershgorin's circle theorem and derive delay-independent results. It is however not clear how delay-dependent results can be obtained, e.g. using delay-dependent convex sets similar to those in Chapter 3.

Most publications on MAS without self-delay rely on sums of Lyapunov-Krasovskii functionals or a contraction argument, see Moreau (2004) for a nice comparison. Sums of Lyapunov-Krasovskii functionals consist of arbitrarily large but finite sums of positive definite terms, each of them related to one agent or one interconnection link. This approach has been used to investigate single integrator MAS (Chellaboina et al., 2006; Ghabcheloo et al., 2007; Papachristodoulou and Jadbabaie, 2005), Internet congestion control (Papachristodoulou and Peet, 2008; Papachristodoulou and Jadbabaie, 2010), and MAS consisting of passive agents or nonlinear agents with relative degree one (Chopra and Spong, 2008, 2006; Chopra et al., 2008). The main disadvantage of this method is that the underlying graph has to be undirected or at least balanced, i.e. the number of incoming and outgoing links of each node are the same. More general results for MAS with single integrator agent dynamics are obtained using a contraction argument (Papachristodoulou and Jadbabaie, 2006; Papachristodoulou et al., 2010) because they only require directed, uniformly quasi-strongly connected graphs, the weakest possible connectivity assumption allowing for consensus. In contrast to Papachristodoulou and Jadbabaie (2006) and our previous publications Münz et al. (2009b,d, 2008a); Papachristodoulou et al. (2010), we use a completely different proof technique in Chapter 4 that makes some technical assumptions obsolete, see Papachristodoulou et al. (2010) for details. Moreover, we consider the more general case of MAS consisting of non-scalar integrators or relative degree one agents with integrating behavior. The presented theorems extend previous results on undelayed single integrator MAS in Lin (2006); Lin et al. (2007) to single integrator MAS with arbitrarily large delays. In contrast to previous publications, we use sums of Lyapunov-Krasovskii functionals for MAS with relative degree two agents in Chapter 5. For these MAS, the contraction argument is not applicable, and the delay robustness of these MAS has not been investigated so far.

1.4 Outline of the Thesis

The thesis starts with a detailed problem statement in Chapter 2, including a motivation for the different delay models. The robustness analysis based on the generalized Nyquist criterion is presented in Chapter 3. Rendezvous in MAS with nonlinear RD1 agents is investigated in Chapter 4. The same problem in MAS with nonlinear RD2 agents is studied in Chapter 5. The thesis is concluded in Chapter 6. Appendix A and B provide introductions to functional differential equations and graph theory. Technical proofs are summarized in Appendix C.

Chapter 2

Problem Statement

In this chapter, we describe the MAS and the cooperative control tasks considered in this thesis. We first present the multi-agent system dynamics in Section 2.1. Then, we explain the different parts of these dynamics. In Section 2.2, we describe the agent dynamics. The interconnection network between the agents is introduced in Section 2.3 with a focus on the network properties topology and delay. The coupling between the agents is explained in Section 2.4. In Section 2.5, we define different cooperative control tasks, namely consensus, rendezvous, flocking, and synchronization. Finally, the different delay models considered in this thesis are motivated and explained in Section 2.6. This chapter is summarized in Section 2.7.

2.1 Multi-Agent System Dynamics

A multi-agent system (MAS) consists of agents $\Sigma_i, i \in \mathcal{N} = \{1, \dots, N\}$ that are coupled through their outputs in order to achieve a cooperative behavior, see Figure 2.1. In this work, we consider MAS with agent dynamics

$$\Sigma_i : \begin{cases} \dot{x}_i(t) = f_i(x_i(t), u_i(t)) \\ y_i(t) = h_i(x_i(t)) \end{cases}, \qquad i \in \mathcal{N} = \{1, \dots, N\}, \tag{2.1a}$$

and the interconnection

$$u_i(t) = -\sum_{j=1}^{N} \frac{a_{ij}}{d_i} k_{ij} \left(\overline{\mathbb{T}}_{ij}(y_{i,t}) - \mathbb{T}_{ij}(y_{j,t}) \right), \tag{2.1b}$$

where $x_i(t) \in \mathbb{R}^{n_i}, u_i(t) \in \mathbb{R}^m, y_i(t) \in \mathbb{R}^m$ are the state, input, and output of agent i, respectively. The functions $f_i : \mathbb{R}^{n_i} \times \mathbb{R}^m \to \mathbb{R}^{n_i}$ and $h_i : \mathbb{R}^{n_i} \to \mathbb{R}^m$ describe the agent dynamics. The topology of the network between the agents is described by the elements of the adjacency matrix of the underlying graph $A = [a_{ij}] \in \mathbb{R}^{N \times N}$, where $a_{ij} > 0$ if there is a link from agent j to agent i are connected and $a_{ij} = 0$ otherwise. The *degree* of agent i is $d_i = \sum_{j=1}^{N} a_{ij}$. The coupling functions $k_{ij} : \mathbb{R}^m \to \mathbb{R}^m$ describe the portion of the input u_i of agent i because of its interaction with agent j. Its argument is the difference between the delayed outputs of agent i and agent j.

The delay operators $\overline{\mathbb{T}}_{ij} : \mathcal{C}_m \to \mathbb{R}^m$ and $\mathbb{T}_{ij} : \mathcal{C}_m \to \mathbb{R}^m$ describe the self-delay and the neighbouring delay of the link from agent j to i, where $\mathcal{C}_m = \mathcal{C}([-\mathcal{T}, 0], \mathbb{R}^m)$ is

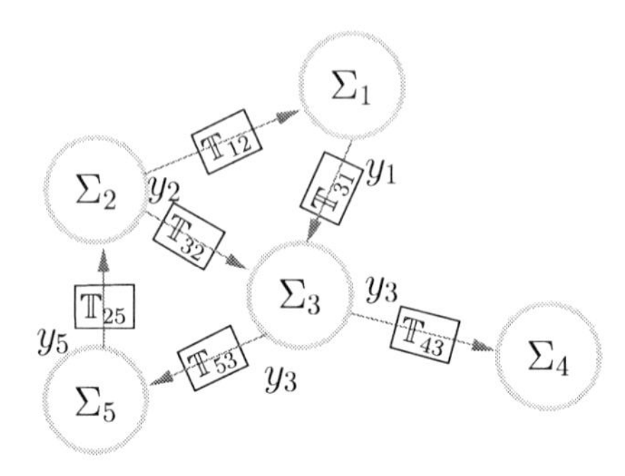

Figure 2.1: Exemplary multi-agent system network with $N = 5$ agents and delays.

the Banach space of continuous functions mapping the interval $[-\mathcal{T}, 0]$ into $\mathbb{R}^m$ for a given *delay range* $\mathcal{T}$, see Appendix A for details. The argument of $\mathbb{T}_{ij}$ is $y_{j,t} \in \mathcal{C}_m$, which denotes the segment of length $\mathcal{T}$ of current and past outputs of agent j. Particular values of the segment $y_{j,t}$ are given as $y_{j,t}(\eta) = y_j(t+\eta)$ for any $\eta \in [-\mathcal{T}, 0]$. The delay operators represent different delay models like constant, time-varying, or distributed delays. For instance, if the channel introduces a constant delay τ_{ij}, then $\mathbb{T}_{ij}(y_{j,t}) = y_j(t - \tau_{ij})$. The delay operators $\overline{\mathbb{T}}_{ij}, \mathbb{T}_{ij}$ are associated to the link from agent j to agent i, see Figure 2.1 where we only depict $\mathbb{T}_{ij}$.

The structure of the MAS (2.1) is summarized in Figure 2.2, which shows a single agent with dynamics (2.1a) and its interconnection to the other agents, depending on the topology and delay of the network as well as on the coupling functions k_{ij}. The closed loop system (2.1) is a retarded functional differential equation (RFDE), see Appendix A for an introduction to RFDEs. Its initial condition is $\varphi \in \mathcal{C}_{\sum n_i} = \mathcal{C}([-\mathcal{T}, 0], \mathbb{R}^{\sum n_i})$. Throughout this thesis, we assume that φ is continuous.

In the following sections, we describe the individual parts of the MAS in detail.

2.2 Agent Dynamics

In this work, we consider MAS that consist of agents with dynamics (2.1a) where the functions f_i and h_i are sufficiently smooth to guarantee existence, uniqueness, and continuity of solutions of the closed loop system (2.1) for sufficiently smooth coupling functions k_{ij}, see Hale and Lunel (1993). In general, the agents can be non-identical, i.e. with different dynamics f_i, h_i and different number of states n_i. Yet, all agents have the same number of inputs and outputs m. It is in fact necessary to assume that all of them have the same number of outputs if we consider the standard cooperative control tasks where eventually $y_i = y_j$ for all i, j, see Section 2.5 below. In addition, it is a natural assumption for MAS that each agent has the same number of inputs and outputs because we want to steer all elements of y_i independently by the control input u_i.

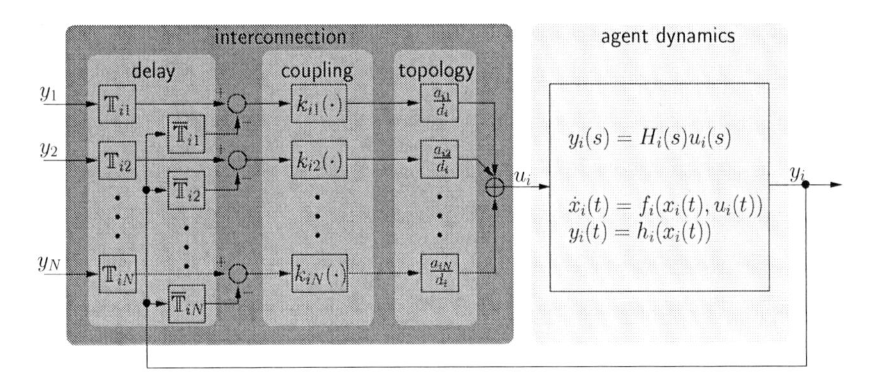

Figure 2.2: Agent i with dynamics (2.1a) and corresponding interconnection (2.1b) depending on the network's topology and delay as well as on the coupling functions k_{ij}.

This assumption excludes, for example, underactuated Euler-Lagrange MAS (Dong and Farrell, 2009; Ortega et al., 1998).

The differential equation (2.1a) contains most agent dynamics considered in the literature as special cases, e.g. single-integrator and double-integrator agent dynamics (Tanner et al., 2007; Jadbabaie et al., 2003; Olfati-Saber et al., 2007; Olfati-Saber, 2006; Ren and Beard, 2008) as well as linear and nonlinear higher order dynamics (Fax and Murray, 2004; Arcak, 2007; Chopra and Spong, 2006; Wieland et al., 2010, 2008).

In Chapter 3, we concentrate on MAS with linear single-input single-output (SISO) agent dynamics with transfer function $H_i(s)$, i.e.

$$\Sigma_i : \quad Y_i(s) = H_i(s)U_i(s), \qquad i \in \mathcal{N} = \{1, \ldots, N\}, \tag{2.2}$$

where $U_i(s)$ and $Y_i(s)$ are the Laplace transform of $u_i(t) \in \mathbb{R}$ and $y_i(t) \in \mathbb{R}$, respectively, see Figure 2.2. Chapter 4 provides results for MAS consisting of nonlinear agents (2.1a) with relative degree one and integrating behavior. In Chapter 5, we investigate a class of nonlinear agents (2.1a) with relative degree two.

2.3 Multi-Agent System Networks

The agents are interconnected on a *network*. Examples for a network are digital packet-switched communication networks (PSCN), power grids, or measurement-based networks as in multi-vehicle applications, where the network links indicate which distances between the vehicles can be measured. In this thesis, we reduce the network to two fundamental properties: the topology and the communication or coupling delays, see Figure 2.1 for an illustration.

We describe the *topology* of the network by a graph $\mathcal{G}(\mathcal{V},\mathcal{E})$ with vertex set $\mathcal{V} = \{v_i\}, i \in \mathcal{N} = \{1,\dots,N\}$, representing the agents and an edge set $\mathcal{E} \subseteq \mathcal{V}\times\mathcal{V}$ representing the links between the agents. The adjacency matrix of the graph is $A = [a_{ij}] \in \mathbb{R}^{N\times N}$. We have $a_{ij} > 0$ if agent i receives information from agent j, i.e. u_i depends on $y_{j,t}$, and $a_{ij} = 0$ otherwise, see also (2.1b). For example, a possible adjacency matrix of the graph in Figure 2.1 is

$$A = \begin{bmatrix} 0 & 2 & 0 & 0 & 0 \\ 0 & 0 & 0 & 0 & 4 \\ 1 & 3 & 0 & 0 & 0 \\ 0 & 0 & 1 & 0 & 0 \\ 0 & 0 & 2 & 0 & 0 \end{bmatrix}. \tag{2.3}$$

The *degree* of agent i is $d_i = \sum_{j=1}^{N} a_{ij}$. Note that d_i is in fact the in-degree of agent i, but we just call it degree for simplicity. The *degree matrix* is thus defined as $D = \operatorname{diag}(d_i)$ and the *Laplacian matrix* of the graph is $L = D - A$. We also define the *normalized Laplacian matrix* $\overline{L} = D^{-1}L = I - D^{-1}A$. Throughout this thesis, we assume that the graph does not have self-loops, i.e. $a_{ii} \neq 0$. The results in Chapter 3 and 5 apply for constant, *undirected* graphs with symmetric weights, i.e. $a_{ij} = a_{ji}$ and $A = A^T$. The results in Chapter 4 also hold for switching, *directed* graphs, i.e. $a_{ij} = a_{ji}$ is not required and the adjacency matrices $A(t)$ is time-varying.

If $a_{ij} > 0$, then agent j is called *parent* of agent i. In undirected graphs, we call agent i and j *neighbours* if $a_{ij} = a_{ji} > 0$. We briefly recall the fundamental connectivity concepts for directed and undirected graphs. An undirected graph is *connected* if any two nodes v_i, v_j of the graph are connected by a path, i.e. a sequence of neighbouring nodes starting at v_i and ending at v_j. A directed graph is *quasi-strongly connected* if there exists a vertex v_R, called the *root*, such that there exist directed paths from v_R to all other nodes v_i of the graph. A *directed path* from v_i to v_j is a sequence of edges out of $\mathcal{E}$ that takes the following form (v_i, v_{i_1}), $(v_{i_1}, v_{i_2}),\dots,(v_{i_p}, v_j)$. Recall that a graph is quasi-strongly connected if and only if it contains a *spanning tree*. A more detailed introduction to graph theory is given in Appendix B.

If $a_{ij} > 0$, agent i receives information about agent j using a link (v_j, v_i). All links of the network are affected by *delays*. These delays can originate from different sources such as transmission delays, scheduling delays, or measurement delays. A more detailed discussion of different kinds of delays is given in Section 2.6 below. The delay from agent j to agent i is modeled using the *delay operator* $\mathbb{T}_{ij} : \mathcal{C}_m \to \mathbb{R}^m$. Hence, the signal at the receiver node v_i at time t is $\mathbb{T}_{ij}(y_{j,t})$ where $y_j(t)$ is the signal at the sender node v_j at time t. Remember that $y_{j,t} \in \mathcal{C}_m$ denotes the segment of length $\mathcal{T}$ of current and past outputs of agent j. Note that $\mathbb{T}_{ij}$ is not a function but a functional on the segment $y_{j,t} \in \mathcal{C}_m$. A more detailed introduction to functionals and functional differential equations is provided in Appendix A.

2.4 Output Interconnection on Networks with Delays

So far, we have established the agent dynamics as well as the network properties topology and delay. It remains to explain the coupling between the agents (2.1b). In Chapter 3, we consider the following delay configurations

$$\text{without self-delay} \qquad u_i(t) = -\sum_{j=1}^{N} \frac{a_{ij}}{d_i}(y_i(t) - \mathbb{T}_{ij}(y_{j,t})), \tag{2.4a}$$

$$\text{with identical self-delay} \qquad u_i(t) = -\sum_{j=1}^{N} \frac{a_{ij}}{d_i}(\mathbb{T}_{ij}(y_{i,t}) - \mathbb{T}_{ij}(y_{j,t})) \tag{2.4b}$$

$$\text{with different self-delay} \qquad u_i(t) = -\sum_{j=1}^{N} \frac{a_{ij}}{d_i}\left(\overline{\mathbb{T}}_{ij}(y_{i,t}) - \mathbb{T}_{ij}(y_{j,t})\right), \tag{2.4c}$$

where the coupling functions are $k_{ij}(\eta) = \eta$ and the delays affect either only the neighbouring output $y_{j,t}$ (2.4a) or also the agent's own output $y_{i,t}$ (2.4b), (2.4c).

The first configuration (2.4a) is denoted *without self-delay* because agent i's own output y_i is not delayed. This configuration models, for example, communication delays for data sent from agent j to agent i, see Section 2.6 for a more detailed motivation of this delay configuration. It has been investigated previously for example in Lee and Spong (2006); Liu and Passino (2006); Papachristodoulou and Jadbabaie (2006, 2005); Chopra and Spong (2006, 2008); Yang et al. (2008).

The second configuration (2.4b) is called *with identical self-delay* because the delay of the output of agent i and of the output of its neighbour j are identical. This configuration is usually proposed for MAS with coupling delays where the delays affect the difference of the agents' outputs $y_{i,t} - y_{j,t}$, e.g. Bliman and Ferrari-Trecate (2008); Xiao and Wang (2008a); Liu and Tian (2008); Tian and Liu (2008); Lestas and Vinnicombe (2007b); Xiao and Wang (2006). Another motivation for (2.4b) are agents with computation or measurement delays, assuming that both the agent's own output and the neighbouring output are affected by the same delay. This has been proposed, for instance, for traffic flow models, see Sipahi and Niculescu (2007); Brackstone and McDonald (1999); Orosz (2005) and references therein.

The third configuration (2.4c) is denoted *with different self-delay* because the delay of the output of agent i and of the output of agent j are different. Motivating applications include different reaction delays to the agent's own behavior and the behavior of its neighbours or computation delays in combination with communication delays, e.g. Xiao and Wang (2008a); Tian and Liu (2008).

A major contribution of Chapter 3 is the analytical comparison of the three delay configurations (2.4) in a unified framework. It turns out that the configuration without self-delay (2.4a) is in general more robust to delays than the other configurations (2.4b),

(2.4c). Note that (2.4a) can be written in the following form

$$u_i(t) = -\sum_{j=1}^{N} \frac{a_{ij}}{d_i} \left(y_i(t) - \mathbb{T}_{ij}(y_{j,t})\right) = -y_i(t) + \sum_{j=1}^{N} \frac{a_{ij}}{d_i} \mathbb{T}_{ij}(y_{j,t}), \tag{2.5}$$

i.e. the control input is the error between the current output $y_i(t)$ of agent i and a weighted average of the delayed outputs of its parents. If the coupling functions k_{ij} are chosen appropriately, the control input of the nonlinear generalization of (2.4a)

$$u_i(t) = -\sum_{j=1}^{N} \frac{a_{ij}}{d_i} k_{ij} \left(y_i(t) - \mathbb{T}_{ij}(y_{j,t})\right), \tag{2.6}$$

which is investigated in Chapter 4 and 5, also corresponds to the error between the agent's own output and a weighted average of the delayed outputs of its parents. This property guarantees that the output y_i moves toward the delayed outputs of its parents if the agent dynamics (2.1a) have an integrating behavior, which is exploited in Chapter 4. This behavior is very intuitive: if people try to meet with their friends, they move toward them or their last known position.

The interconnection (2.6) is fairly simple because it is static and does not require to estimate the states of the neighbours. The possibly nonlinear functions k_{ij} are such that different nonlinearities may affect each link. This is particularly useful for local, scalable design conditions as presented in Chapter 5, where particular bounds on k_{ij} depend on the delays $\mathbb{T}_{ij}$ of the corresponding channel. Finally, (2.6) includes typical feedback interconnections in the literature as special cases, e.g. the interconnection of Kuramoto oscillators (Strogatz, 2000; Kuramoto, 1984), of car-following models (Helbing, 2001), and in power networks (Pavella and Murthy, 1994).

2.5 Cooperative Control Tasks

The aim of the MAS (2.1) is to achieve a cooperative behavior of the agents. We call this aim a *cooperative control task*. There are many different cooperative control applications as described in the Introduction. In this work, we focus on the most fundamental tasks consensus, rendezvous, flocking, and synchronization. These terms are not used consistently in the literature. Throughout this thesis, we rely on the following definitions: *Consensus* describes a behavior where agents achieve an agreement on a scalar value. *Rendezvous* refers to agents arriving at the same point in a higher dimensional space, e.g. underwater vehicles that meet somewhere in the sea. *Flocking* is a more complex, swarm like behavior where all agents eventually move in the same direction, i.e. they achieve an agreement on the velocity, and end up in a particular formation. This formation problem is often simplified to all agents being at the same point and moving in the same direction. Here, we also consider this simplified flocking problem. *Synchronization* refers to a set of oscillators that eventually oscillate with the same frequency and phase. A formal definition of consensus, rendezvous, flocking, and synchronization is given next. Thereby, $\|\cdot\|$ denotes the Euclidean norm.

Definition 2.1 (Consensus). *A MAS consisting of agents* (2.1a) *with* $y_i(t) \in \mathbb{R}, u_i(t) \in \mathbb{R}$, *i.e.* $m = 1$, *achieves a* consensus *asymptotically if*

$$\begin{aligned} \lim_{t\to\infty} (y_i(t) - y_j(t)) &= 0, && \text{for all } i, j, \\ \lim_{t\to\infty} \dot{y}_i(t) &= 0, && \text{for all } i. \end{aligned} \tag{2.7}$$

Definition 2.2 (Rendezvous). *A MAS consisting of agents* (2.1a) *with* $y_i(t) \in \mathbb{R}^m$, $u_i(t) \in \mathbb{R}^m$, $m \geq 1$, *achieves a* rendezvous *asymptotically if*

$$\begin{aligned} \lim_{t\to\infty} \|y_i(t) - y_j(t)\| &= 0, && \text{for all } i, j, \\ \lim_{t\to\infty} \dot{y}_i(t) &= 0, && \text{for all } i, \end{aligned} \tag{2.8}$$

where $\|\cdot\|$ *is the Euclidean vector norm.*

Definition 2.3 (Flocking). *A MAS consisting of agents* (2.1a) *achieves* flocking *asymptotically if*

$$\begin{aligned} \lim_{t\to\infty} \|y_i(t) - y_j(t)\| &= 0, && \text{for all } i, j, \\ \lim_{t\to\infty} \|\dot{y}_i(t) - c(t)\| &= 0, && \text{for all } i, \end{aligned} \tag{2.9}$$

for any in general nonzero $c : \mathbb{R} \to \mathbb{R}^m$.

Definition 2.4 (Synchronization). *A MAS consisting of agents* (2.1a) *achieves* synchronization *asymptotically if*

$$\begin{aligned} \lim_{t\to\infty} \|y_i(t) - y_j(t)\| &= 0, && \text{for all } i, j, \\ \lim_{t\to\infty} \|\dot{y}_i(t) - c(t)\| &= 0, && \text{for all } i, \end{aligned} \tag{2.10}$$

where $c : \mathbb{R} \to \mathbb{R}^m$ *is the derivative of the periodic synchronized orbit of* y_i *with period* $T_p > 0$, *i.e.* $\int_t^{t+T_p} c(\eta) d\eta = 0$ *for all* t.

Note that these definitions are not unambiguous, e.g. flocking includes synchronization as special case. However, comparing Definition 2.1 and 2.2 to Definition 2.3 and 2.4, we see that consensus and rendezvous describe a static terminal state, i.e. all outputs are constant as soon as consensus or rendezvous is reached. On the contrary, flocking and synchronization denote a dynamic terminal state where the agents still move when flocking or synchronization is achieved.

It is worth mentioning that there is a fundamental difference between standard stability problems and cooperative control problems like consensus and rendezvous. Standard stability problems analyse the asymptotic stability of isolated steady states, i.e. the goal is to prove that all trajectories in the neighbourhood of this steady state converge toward this steady state. Rendezvous and consensus problems usually exhibit unbounded, connected sets of steady states. The challenge is to prove that all trajectories starting in the neighbourhood of this set of steady states asymptotically converge toward this set, i.e. the set of steady states is asymptotically attracting. Note however that the individual steady states in this set are not asymptotically stable because in every arbitrarily small neighbourhood of any element of this set are further steady states. The exact point in the set, where the system converges to, is in many cases irrelevant.

In the literature on MAS with constant delays τ_{ij}, an alternative condition for rendezvous and flocking has been proposed:

$$\lim_{t\to\infty} \|y_i(t) - y_j(t-\tau_{ij})\| = 0, \quad \forall i,j, \tag{2.11}$$

e.g. Chopra and Spong (2008, 2006). For rendezvous, this condition is equivalent to the condition (2.8) because $y_j(t-\tau_{ij}) = y_j(t)$ if $\dot{y}_j(t) \equiv 0$. Yet, condition (2.11) is only suitable for dynamic cooperative control tasks under additional assumptions on the delays, which is illustrated in the following example:

Example 2.5. *Consider a MAS consisting of two agents with scalar output $y_i(t)$. If flocking is reached according to (2.11), then the outputs eventually satisfy*

$$\begin{aligned} y_1(t) &= \tilde{c}(t) + y_{10} \\ y_2(t) &= \tilde{c}(t) + y_{20}, \end{aligned} \tag{2.12}$$

for some $\tilde{c}(t) = \int_0^t c(\eta)d\eta$ with $c(t)$ as in (2.9). Solving (2.11) with (2.12) gives

$$\begin{aligned} y_1(t) - y_2(t-\tau_{21}) = 0 = \tilde{c}(t) - \tilde{c}(t-\tau_{21}) + y_{10} - y_{20} \\ y_2(t) - y_1(t-\tau_{12}) = 0 = \tilde{c}(t) - \tilde{c}(t-\tau_{12}) + y_{20} - y_{10}, \end{aligned}$$

and combining these two equations, we see that flocking requires

$$2\tilde{c}(t) - \tilde{c}(t-\tau_{21}) - \tilde{c}(t-\tau_{12}) = 0. \tag{2.13}$$

For flocking with constant derivatives $c(t) = c_0 \neq 0$, i.e. $\tilde{c}(t) = c_0 t$, Equation (2.13) holds only for zero delays $\tau_{21} = \tau_{12} = 0$.

For non-constant derivatives $c(t) \neq$ const., Equation (2.13) holds, e.g., for periodic functions $c(t) = c(t-\tau_{21}) = c(t-\tau_{12})$ such as $c(t) = \sin(\omega_0 t), \omega_0 > 0$, if $\tau_{21} = \frac{2\pi n_1}{\omega_0}, \tau_{12} = \frac{2\pi n_2}{\omega_0}, n_1, n_2 \in \mathbb{N}$. Note however that these periodic solutions only exist in very particular cases with suitable agent dynamics, e.g. harmonic oscillators, and perfectly tuned delays, i.e. these solutions are not robust to uncertain delays.

This simple example shows that condition (2.11) imposes additional assumptions on the delays if dynamic cooperative control tasks are considered. These assumptions are even more complicated if large MAS with many heterogeneous delays are studied. Therefore, Definitions 2.3 and 2.4 are preferable.

In Chapter 3, we mainly focus on consensus problems. We also show that flocking in MAS with delays is only possible in particular configurations. Chapter 4 and 5 provide solutions to consensus and rendezvous problems with application examples on synchronization in Section 4.4 and on flocking in Section 5.3.2.

2.6 Delay Models

Delays appear in many interconnection networks. In "biological" MAS like schools of fish or car-following scenarios, we often encounter *reaction delays* that stem from the reaction time of the biological agents with respect to the movement of their fellow agents. Reaction delays are typically in the range of several hundreds of milliseconds. We know from the literature on car-following that reaction delays are important in this area for accurate modeling (Brackstone and McDonald, 1999; Orosz, 2005). In man-made networks like power networks, *propagation delays* are induced because of the long links between different power suppliers and consumers. Simplifying models reduce these propagation delays to complex admittances at the driving frequency 50Hz or 60Hz (Pavella and Murthy, 1994). Finally, digital PSCN introduce *queuing delays* and *access delays*. Access delays emerge because many digital PSCN use a shared communication medium. Hence, information cannot be transmitted instantaneously from node i to node j but only if node i has access to the communication medium. In some PSCN, access is given to the nodes by scheduling algorithms like token-passing, which induces delays because node i has to wait for the token before it may transmit data. Other PSCN allow for random access which might result in collisions. The necessary collision arbitration or avoidance procedures usually induce access delays in the range of at least several hundreds of microseconds even in small networks, see Moyne and Tilbury (2007). Queuing delays result from the queuing of packets in routers of PSCN. Both access and queuing delays are also usually randomly distributed depending on the network load.

We see that delays are an important characteristic of networks. In the MAS literature, delays are often neglected because they are assumed to be small. Yet, we have seen that reaction delays in biological MAS are usually around several hundreds of milliseconds, and in some cases, they are essential for an accurate modeling. Moreover, if access and queuing delays are in the range of hundreds of microseconds, we may only neglect them if we are sure that the MAS is robust to these delays. Hence, the important question arises: When are these delays sufficiently small such that they do not affect the cooperative control task? The answer to this question is given by the delay robustness analysis in this thesis.

This thesis is focused on communication and coupling delays between the agents. There is another important source of delays in MAS that we consider here only as special cases of the different feedback delay configurations (2.4): *computation delays*, i.e. the time an agent needs to compute the control input u_i from the available outputs $y_i(t)$ and $y_j(t)$. An interesting discussion on the range and influence of computation and communication delays in industrial PSCN is given in Moyne and Tilbury (2007).

In this thesis, we consider three different kinds of delays, namely constant delays, time-varying delays, and distributed delays, in order to model different sources of delays. These delays are introduced in the following:

Constant delays The simplest delay model for a link between two agents is a *constant delay* $\tau_{ij} \in [0, \mathcal{T}]$ with corresponding delay operator

$$\mathbb{T}_{ij}(y_{j,t}) = y_j(t - \tau_{ij}). \tag{2.14}$$

Constant delays are usually a first step to investigate the robustness of a MAS to delays. They are an accurate model for reaction and propagation delays in some cases. For reaction delays, we have to assume that the reaction time is not changing, e.g., due to changing concentration on the cooperative control task. For propagation delays, we have to assume that the distance between the nodes i and j remains constant. This might contradict the idea of consensus and rendezvous if $y_j(t) - y_i(t)$ describes the distance between agent i and j. In this case, we expect that the propagation delay decreases as rendezvous is approached. Constant delays are in general unsuited for access delays, which are usually random delays depending, e.g., on the network load.

Time-varying delays An appropriate generalization of constant delays, e.g. for random access delays, are *time-varying delays* $\tau_{ij} : \mathbb{R} \to [0, \mathcal{T}_{ij}]$, with $\max_{i,j} \mathcal{T}_{ij} = \mathcal{T}$ and delay operator

$$\mathbb{T}_{ij}(y_{j,t}) = y_j(t - \tau_{ij}(t)). \tag{2.15}$$

Throughout this thesis, we consider deterministic, piecewise continuous time-varying delays τ_{ij} such that discontinuities in τ_{ij} are separated by a dwell time $h_{\tau_{ij}} > 0$, i.e. consecutive discontinuity points $t_{l+1} > t_l$ satisfy $t_{l+1} - t_l \geq h_{\tau_{ij}}$ for all l. This delay model covers reaction, propagation, queuing, and access delays equally well. Yet, its major drawback is conservativity if some information about the evolution of τ_{ij} is known. For example, in PSCN with random access, the majority of the data is transmitted with a rather small delay whereas very few data is heavily delayed with delays close to the delay bound $\mathcal{T}_{ij}$ (Moyne and Tilbury, 2007). Intuitively, we expect that the consensus properties of a MAS using this network depend rather on the usual small delays than on the seldom large delays. However, delay-dependent stability and consensus conditions for systems with time-varying delays usually depend only on the maximal delay $\mathcal{T}_{ij}$. Since larger delays usually deteriorate consensus properties, time-varying delays often lead to conservative conditions for rendezvous. This is shown in an example in Section 5.4.3.

Distributed delays A compromise between accurate time-varying delay models and less conservative consensus conditions are *distributed delays*, where the delay operator is

$$\mathbb{T}_{ij}(y_{j,t}) = \int_0^{\mathcal{T}_{ij}} \phi_{ij}(\eta) y_j(t - \eta) d\eta. \tag{2.16}$$

The output $\mathbb{T}_{ij}(y_{j,t})$ of a distributed delay channel depends on the weighted average of the input $y_j(\eta)$ in the interval $\eta \in [t - \mathcal{T}_{ij}, t]$. The weighting function is the *delay kernel* $\phi_{ij} : \mathbb{R} \to \mathbb{R}$ that satisfies $\phi_{ij}(\eta) \geq 0$ for all $\eta \in [0, \mathcal{T}_{ij}]$ and $\int_0^{\mathcal{T}_{ij}} \phi_{ij}(\eta) d\eta = 1$. It has been shown in Michiels et al. (2005) that fast time-varying delays can be approximated by distributed delay with an appropriate kernel ϕ_{ij}. Moreover, the resulting distributed delay model may lead to less conservative stability conditions than the time-varying delay model, e.g. Michiels et al. (2005). We have shown in Münz et al. (2009b) that distributed delays are also a suitable model for continuous streams of packets in PSCN. In this context, $\phi_{ij}(\eta)$ describes the probability density that a packet is delayed by η, which in turn implies $\int_0^{\mathcal{T}_{ij}} \phi_{ij}(\eta)d\eta = 1$. Similarly, distributed delays are often used to model stochastic delays, e.g. the reproduction delay in population dynamics (Cushing, 1977). We will show in Chapter 5 that delay-dependent rendezvous conditions for MAS on PSCN modeled with distributed delays depend on the average delay $\overline{\tau}_{ij} = \int_0^{\mathcal{T}_{ij}} \eta\phi_{ij}(\eta)d\eta$ of the communication link. Yet, the same condition depends on the maximal delay $\mathcal{T}_{ij}$ if the PSCN is modeled with time-varying delays. Hence, distributed delay models may lead to less conservative conditions than time-varying delay models, see also the example in Section 5.4.3.

Distributed delays (2.16) can also be interpreted as a convolution between ϕ_{ij} and the elements of y_j. Thus, they also describe networks with general linear, time-invariant (LTI) channel dynamics. This interpretation is particularly suited for propagation delays, e.g. in power networks. In this context, ϕ_{ij} is the impulse response of the channel.

Modeling delays in networks This short presentation of the different delay models already indicated for what kind of delays these models are adequate. In general, the analysis based on constant delays is suitable if the delays are indeed constant or only slowly time-varying. This is however not the case in many MAS with delays where the delays are often randomly time-varying. For these MAS, delay robustness analysis for constant delays can be used as a first indicator about the delay robustness of the MAS with time-varying delays because constant delays are a special case of time-varying delays. Yet, a correct delay robustness analysis can only be obtained using time-varying delay models. However, time-varying delays lead often to quite conservative results. Therefore, distributed delays are preferable if consecutive time-delays are uncorrelated and if the distribution of the time-varying delays is known. In these cases, the delay robustness analysis based on distributed delay models is usually less conservative than the analysis based on time-varying delays. In view of these advantages and disadvantages of the different delay models, it is a major advantages of the results presented in Chapter 4 and 5 that they apply equally to all three delay models. Especially, the conditions in Chapter 5 show the influence of the different delay models on the delay robustness.

Most publications on stability analysis and control of time-delay systems consider constant delays, see for example (Gu, 2003; Jarlebring, 2008; Gouaisbaut and Peaucelle, 2006; Münz et al., 2009a, 2007a; Haag et al., 2009; Maier et al., 2010; Mahboobi Esfanjani et al., 2009), less results are available for time-varying and distributed delays, e.g. (Fridman et al., 2004; Mehdi et al., 2002; Gu et al., 2003; Morărescu et al., 2007; Michiels et al., 2005; Münz et al., 2009f, 2008c; Münz and Allgöwer, 2007; Goebel et al., 2010), see also Niculescu (2001); Gu et al. (2003); Richard (2003); Michiels and Niculescu (2007); Loiseau et al. (2009) and references therein. Considering the literature on MAS, there are only few publications on delay robustness. A detailed overview has been given in the Introduction. Most publications deal with constant delay models, e.g. Olfati-Saber and Murray (2004); Moreau (2004); Papachristodoulou and Jadbabaie (2006); Chopra and Spong (2006). Time-varying and distributed delays are only considered in very rare publications (Bliman and Ferrari-Trecate, 2008; Yang et al., 2008; Michiels et al., 2009). MAS with general LTI channels have been considered in Wang and Elia (2008); Lestas and Vinnicombe (2007b).

2.7 Summary

In this chapter, we have introduced the setup of the cooperative control tasks considered in this thesis. We presented all parts of the multi-agent system: the agent dynamics, the network topology, the feedback interconnection, and different cooperative control tasks. Moreover, we provided a detailed motivation and description of the different delay models considered in this thesis.

Chapter 3

Linear Multi-Agent Systems

In this chapter, we develop a method to analyse the delay robustness of consensus and flocking algorithms in linear multi-agent systems (MAS). We investigate and compare different feedback delay configurations. Using this method, we study the delay robustness of several special cases, like single integrator MAS.

We start with an introduction of the agent dynamics and different feedback delay configurations in Section 3.1. In Section 3.2, we recall consensus conditions for the undelayed MAS. Then, we investigate if the different feedback delay configurations have consensus and flocking solutions in Section 3.3. The main result of this chapter is presented in Section 3.4: a method to analyse the delay robustness of consensus and flocking in linear MAS. The following sections consider special cases of linear MAS: In Section 3.5, we apply this method to single integrator MAS. In Section 3.6, we investigate the delay robustness of a class of relative degree two MAS. A consensus controller design algorithm is presented in Section 3.7. Finally, we analyse the convergence rate of linear single integrator MAS in Section 3.8. A summary is given in Section 3.9. Preliminary results of this chapter have been published in Münz et al. (2010c, 2009c,e).

3.1 Agent Dynamics and Delay-Interconnection Matrices

3.1.1 Agent Dynamics

In this chapter, we consider MAS consisting of N linear, single-input, single-output (SISO) agents described by strictly proper transfer functions $H_i(s)$

$$Y_i(s) = H_i(s)U_i(s) = \frac{\nu_i(s)}{\delta_i(s)}U_i(s), \qquad i \in \mathcal{N} = \{1, 2, \ldots, N\}, \tag{3.1}$$

where $U_i(s)$ and $Y_i(s)$ are the Laplace transform of the input $u_i(t)$ and output $y_i(t)$ of agent i, see (2.1a), (2.2). These agent dynamics include the most typical MAS models like single or double integrator MAS as special case. We assume the following on H_i:

Assumption 3.1. *The numerator polynomials $\nu_i(s) = \nu_{i0} + \nu_{i1}s + \ldots + \nu_{i\tilde{n}_i}s^{\tilde{n}_i}, \nu_{il} \in \mathbb{R}$, for all $l = 1, \ldots, \tilde{n}_i$, and denominator polynomials $\delta_i(s) = \delta_{i0} + \delta_{i1}s + \ldots + \delta_{in_i}s^{n_i}, \delta_{il} \in \mathbb{R}$, for all $l = 1, \ldots, n_i$, of the rational transfer functions $H_i(s) = \frac{\nu_i(s)}{\delta_i(s)}$ satisfy the following:*

- *all polynomials $\delta_i(s), i \in \mathcal{N}$, have a root at the origin, possibly with multiplicity larger than one;*

- *$\nu_i(s)$ and $\delta_i(s)$ are coprime and the transfer functions $H_i(s)$ are strictly proper, i.e. $n_i > \tilde{n}_i$; and*

- *$\delta_i(s) + \varrho\nu_i(s)$ is Hurwitz for all $i \in \mathcal{N}$ and for all $\varrho \in (0, 2]$, which implies that all nonzero roots of $\delta_i(s)$ are in the open left half plane $\mathbb{C}^-$.*

We briefly motivate these assumptions. The main focus of this chapter are consensus problems, i.e. the output differences $y_i(t) - y_j(t)$ vanish over time. This is trivially satisfied without any coupling if all agents are asymptotically stable, i.e. all outputs $y_i(t)$ converge to the origin. In order to exclude this trivial consensus problem, we assume that the polynomials $\delta_i(s)$ have a root at the origin. Therefore, the agents are not asymptotically stable. If this zero root is a single root, all outputs $y_i(t)$ eventually approach constant, in general nonzero values if $u_i(t) \equiv 0$, for all $i \in \mathcal{N}$. The cooperative control objective is that all outputs end up at the same value. Similar arguments apply for multiple poles at the origin, see also Section 3.3. The second item in Assumption 3.1 is not restrictive because we consider controllable and observable agent dynamics. In this case, there always exists a transfer function with coprime $\nu_i(s)$ and $\delta_i(s)$. Moreover, we assume that the transfer functions H_i are strictly proper which is necessary in the undelayed case in order to exclude algebraic loops in the output interconnection described below. Finally, the Hurwitz assumption on $\delta_i(s) + \varrho\nu_i(s)$, $\varrho \in (0, 2]$, in the last item guarantees that consensus is achieved in MAS (3.1) without delays and arbitrary topology, which is shown in Lemma 3.2 below. This also implies that all modes of each agent are asymptotically stable, except for those modes with zero eigenvalue.

3.1.2 Delay-Interconnection Matrices

In this chapter, we investigate output consensus and flocking conditions for MAS (3.1) as defined in Definition 2.1 and 2.3, i.e. the aim of the MAS is

$$\lim_{t\to\infty} \|y_i(t) - y_j(t)\| = 0, \quad \text{for all } i, j. \tag{3.2}$$

For simplicity of presentation, we often say consensus in this chapter when referring to consensus or flocking. In order to achieve consensus, the agents exchange their outputs y_i using a network with constant, undirected, connected topology with symmetric weights $a_{ij} = a_{ji}$, i.e. the adjacency matrix of the underlying graph is symmetric $A = A^T$. Since the graph is connected, all agents have at least one neighbour, i.e. the degree of each node is $d_i > 0$. We investigate and compare three cases how delays affect the coupling between the agents. The corresponding interconnections are denoted

without self-delay $$u_i(t) = -\sum_{j=1}^{N} \frac{a_{ij}}{d_i}(y_i(t) - y_j(t-\tau_{ij})), \tag{3.3a}$$

with identical self-delay $$u_i(t) = -\sum_{j=1}^{N} \frac{a_{ij}}{d_i}(y_i(t-\tau_{ij}) - y_j(t-\tau_{ij})), \tag{3.3b}$$

with different self-delay $$u_i(t) = -\sum_{j=1}^{N} \frac{a_{ij}}{d_i}(y_i(t-T_{ij}) - y_j(t-\tau_{ij})), \tag{3.3c}$$

where we focus on constant time-delays. Since time-varying delays cannot be transformed into the Laplace domain, the derived conditions are not applicable for MAS with time-varying delays. We will briefly explain in the Conclusions of this thesis how the results of this chapter can be extended to analyse MAS with distributed delays. We denote MAS (3.1) with interconnection (3.3a) *MAS without self-delay*, with interconnection (3.3b) *MAS with identical self-delay*, and with interconnection (3.3c) as *MAS with different self-delay*. Note that (3.3c) includes (3.3b) for $T_{ij} = \tau_{ij}$ and (3.3a) for $T_{ij} = 0$ as special cases. Later on, we will in fact differentiate between four cases because we obtain different conditions for MAS with symmetric identical self-delay $\tau_{ij} = \tau_{ji}$, for all i, j and with asymmetric identical self-delay $\tau_{ij} \neq \tau_{ji}$ for some i, j.

Now, we can introduce the initial condition of the MAS with agents (3.1) and interconnection (3.3): $\varphi \in \mathcal{C}_{\sum n_i} = \mathcal{C}([-\mathcal{T}, 0], \mathbb{R}^{\sum n_i})$, see also Section 2.1. We assume that φ is continuous.

In order to transform the interconnections (3.3) in the Laplace domain, we define the delay-dependent adjacency and degree matrices

$$A_\tau(s) = \left[a_{ij}e^{-\tau_{ij}s}\right], \quad D_\tau(s) = \operatorname{diag}\left(\sum_{j=1}^{N} a_{ij}e^{-\tau_{ij}s}\right), \quad D_T(s) = \operatorname{diag}\left(\sum_{j=1}^{N} a_{ij}e^{-T_{ij}s}\right). \tag{3.4}$$

Note that A_τ, D_τ, and D_T contain the full information about network properties topology and delay. Now, we can define the *delay-interconnection matrices* $\Gamma_r : \mathbb{C} \to \mathbb{C}^{N\times N}, r = 1, 2, 3, 4$, that result from the Laplace transform of the interconnections (3.3):

without self-delay $$U(s) = -\Gamma_1(s)Y(s) = -(I - D^{-1}A_\tau(s))Y(s), \tag{3.5a}$$

with identical self-delay $$U(s) = -\Gamma_2(s)Y(s) = -D^{-1}L_\tau(s)Y(s), \tag{3.5b}$$

with different self-delay $$U(s) = -\Gamma_3(s)Y(s) = -D^{-1}L_{\tau T}(s)Y(s), \tag{3.5c}$$

with symmetric identical self-delay i.e. $\tau_{ij} = \tau_{ji}$, $$U(s) = -\Gamma_4(s)Y(s) = -D^{-1}L_\tau(s)Y(s), \tag{3.5d}$$

where $U(s) = [U_1(s), \ldots, U_N(s)]^T$, $Y(s) = [Y_1(s), \ldots, Y_N(s)]^T$, $L_\tau(s) = D_\tau(s) - A_\tau(s)$, $L_{\tau T}(s) = D_T(s) - A_\tau(s)$ and $D = \operatorname{diag}(d_i)$. The only difference between Γ_2 and Γ_4 is that Γ_4 requires symmetric delays $\tau_{ij} = \tau_{ji}$ and Γ_2 allows also for asymmetric delays. This distinction will become important later on.

An important feature of the feedback matrices Γ_r is the right eigenvector $\mathbf{1} = [1\,1\dots 1]^T$. Note that

$$\Gamma_r(0)\mathbf{1} = 0 \qquad \text{for } r = 1,2,3,4, \qquad \text{and} \tag{3.6a}$$

$$\Gamma_r(s)\mathbf{1} = 0 \qquad \text{for all } s \in \mathbb{C}, r = 2,4. \tag{3.6b}$$

Yet, $\Gamma_r(s)\mathbf{1} = 0$ for all $s \in \mathbb{C}$ and for $r = 1,3$ is not true in general. This already indicates a fundamental difference between MAS with identical self-delay, i.e. Γ_2, Γ_4, and MAS without or with different self-delay, i.e. Γ_1, Γ_3, which will be further investigated in Section 3.3.

In Section 3.4, we develop a methodology to analyse the consensus properties of general linear MAS (3.1) with interconnections (3.5). The method is based on the generalized Nyquist criterion (Desoer and Wang, 1980) and provides sufficient set-valued conditions. This approach allows for a detailed comparison of the different delay configurations.

3.2 Consensus of Undelayed Multi-Agent Systems

Before we study the consensus problem of MAS with delays, we want to recall the consensus properties of MAS with identical agents and without delay, i.e. with $H_i(s) = H(s) = \frac{\nu(s)}{\delta(s)}$ and interconnection matrix $\Gamma_0 = \overline{L} = I - D^{-1}A$. This MAS corresponds to (3.1) with interconnection (3.5) with zero delay $\tau_{ij} = T_{ij} = 0$ for all i, j. This consensus problem has been studied previously in Fax and Murray (2004); Wieland et al. (2008, 2010); Ren and Beard (2008) and is presented here for completeness. The characteristic polynomial of the closed loop system is

$$\Delta_0(s) = \det\left(\delta(s) + \nu(s)\overline{L}\right) = \prod_{l=1}^{N}\left(\delta(s) + \nu(s)\lambda_l(\overline{L})\right), \tag{3.7}$$

where $\lambda_l(\overline{L})$ are the eigenvalues of $\overline{L}$ such that $0 = \lambda_1(\overline{L}) < \lambda_2(\overline{L}) \leq \dots \leq \lambda_N(\overline{L})$. Since the graph is undirected and connected, all eigenvalues of $\overline{L}$ are real, see Appendix B, and $\lambda_2(\overline{L}) > 0$. The largest eigenvalue of the normalized Laplacian matrix satisfies $\lambda_N(\overline{L}) \leq 2$. This can be shown using Gershgorin's circle theorem, see Horn and Johnson (1985) for example, and noting that all diagonal elements of $\overline{L}$ are 1 and the row elements sum up to zero. The eigenvector of $\overline{L}$ corresponding to $\lambda_1(\overline{L}) = 0$ is $\mathbf{1}$, see Appendix B, which corresponds to the consensus solutions with $y_1(t) = y_2(t) = \dots = y_N(t)$. The attractivity of this consensus solution depends only on the nonzero eigenvalues of $\overline{L}$. Sufficient conditions for reaching consensus asymptotically without delays are provided in the next lemma.

Lemma 3.2 (Consensus of undelayed MAS). *Consider a MAS consisting of agents dynamics (3.1) with identical dynamics $H_i(s) = H(s)$ and symmetric interconnection matrix without delays $\Gamma_0 = \overline{L} = I - D^{-1}A$. This MAS achieves consensus for arbitrary, undirected, connected topologies if $\delta(s) + \varrho\nu(s)$ is Hurwitz for all $\varrho \in (0, 2]$.*

Proof. As we are interested in the convergence toward the consensus set, we only consider the nonzero eigenvalues of $\overline{L}$. Thus, the MAS achieves consensus if and only if

$$\prod_{l=2}^{N} \left(\delta(s) + \nu(s)\lambda_l(\overline{L})\right) \neq 0.$$

for all s with $\mathrm{Re}\{s\} \geq 0$. Since $\lambda_l(\overline{L}) \in (0,2]$, $\delta(s) + \varrho\nu(s), \varrho \in (0,2]$, being Hurwitz guarantees consensus. □

In Lemma 3.2, consensus has to hold for arbitrary large networks with arbitrary topologies. For arbitrary topologies, the nonzero eigenvalues of $\overline{L}$ can take arbitrary values in $(0,2]$. In particular, we can construct topologies such that $\lambda_N(\overline{L}) = 2$, e.g. $\overline{L} = I - D^{-1}A$ with A given in (3.31) below, and other topologies such that $\lambda_2(\overline{L}) \leq \epsilon$ for arbitrary small $\epsilon > 0$, see Example 1.4 and 1.5 in Chung (1997). Lemma 3.2 also illustrates why we include $\delta_i(s) + \varrho\nu_i(s), \varrho \in (0,2]$, to be Hurwitz in Assumption 3.1. Recall that this also implies that $\delta_i(s)$ has all roots in $\mathbb{C}_0^-$.

3.3 Consensus and Flocking Solutions

In this section, we show a fundamental difference between the interconnections (3.3). We will see that MAS without self-delay and MAS with different self-delay in general only have consensus solutions, whereas MAS with identical self-delay also have flocking solutions and achieve *average consensus* and *average flocking.* Average consensus and flocking means that the average of all outputs of the closed loop system $\overline{y}(t) = \frac{1}{N}\sum_{j=1}^{N} y_j(t)$ is a solution of the agent dynamics H with zero input and average initial condition $[\overline{y}(0), \dot{\overline{y}}(0), \ldots, \overline{y}^{(n-1)}(0)]^T$. We consider here again MAS with identical dynamics $H_i(s) = H(s) = \frac{\nu(s)}{\delta(s)}$ in order to simplify the presentation. Basically the same results are obtained for non-identical agent dynamics.

3.3.1 Closed Loop Dynamics with Initial Conditions

We first have a closer look at the agent dynamics including their initial conditions. The transfer function H in (3.1) corresponds to the following minimal realization differential equation

$$\sum_{l=0}^{n} \delta_l y_i^{(l)}(t) = \sum_{l=1}^{\tilde{n}} \nu_l u_i^{(l)}(t),$$

with initial condition $y_i(0), \ldots, y_i^{(n_i-1)}(0), u_i(0), \ldots, u_i^{(\tilde{n}_i-1)}(0)$, where $y_i(t), u_i(t)$ are the inverse Laplace transforms of $Y_i(s), U_i(s)$ and $y_i^{(\iota)}$ is the ι-th derivative of y_i. We write $\delta_{il} = \delta_l$ and $\nu_{il} = \nu_l$ because all H_i are identical. Applying the Laplace transform leads to

$$\delta(s)Y_i(s) - Y_{0,i}(s) = \nu(s)U_i(s) - U_{0,i}(s),$$

$$\text{where} \qquad Y_{0,i}(s) = \sum_{l=1}^{n} \delta_l \sum_{\iota=0}^{l} s^{l-\iota-1} y_i^{(\iota)}(0),$$

$$U_{0,i}(s) = \sum_{l=1}^{\tilde{n}} \nu_l \sum_{\iota=0}^{l} s^{l-\iota-1} u_i^{(\iota)}(0).$$

The coefficients of the polynomials $Y_{0,i}$ and $U_{0,i}$ depend on the initial conditions and the coefficients of $\delta(s)$ and $\nu(s)$, respectively. For zero initial conditions, we recover the agent dynamics (3.1), i.e. $\delta(s)Y_i(s) = \nu(s)U_i(s)$.

Now, consider the interconnections (3.3). With this feedback, the initial conditions of the input $u_i(0), \ldots, u_i^{(\tilde{n}_i-1)}(0)$ can be computed directly from the initial conditions $y_i(\eta), \eta \in [-\mathcal{T}, 0]$, of the outputs and the delays τ_{ij}, T_{ij}. Exemplarily, we compute the most general case (3.3c) as

$$u_i^{(\iota)}(0) = -\sum_{j=1}^{N} \frac{a_{ij}}{d_i} (y_i^{(\iota)}(-T_{ij}) - y_j^{(\iota)}(-\tau_{ij})). \tag{3.8}$$

Additional initial conditions have to be included in the Laplace transform of (3.3) due to the delays. Note that

$$\mathcal{L}(y_j(t-\tau_{ij})) = e^{-\tau_{ij}s} \left(Y_j(s) + \int_{-\tau_{ij}}^{0} y_j(\eta) e^{-s\eta} d\eta \right).$$

It is often assumed that $y_j(\eta) = 0$ for all $\eta < 0$, i.e. the integral in the previous equation is zero. However, we consider here also nonzero initial conditions, i.e. the integral above is not zero. We transform the most general case (3.3c) in the Laplace domain including the initial conditions:

$$U_i(s) = -\sum_{j=1}^{N} \frac{a_{ij}}{d_i} \left(e^{-T_{ij}s} \left(Y_i(s) + \int_{-T_{ij}}^{0} y_i(\eta) e^{-s\eta} d\eta \right) \right.$$
$$\left. - e^{-\tau_{ij}s} \left(Y_j(s) + \int_{-\tau_{ij}}^{0} y_j(\eta) e^{-s\eta} d\eta \right) \right).$$

The Laplace transform of the feedback interconnections (3.3a) and (3.3b) can be derived for particular choices of T_{ij}.

Finally, we can write the closed loop dynamics of the complete MAS including initial conditions in the Laplace domain as

$$\delta(s)Y(s) = -\nu(s)\Gamma_r(s)Y(s) - \nu(s)\mathcal{Y}_0(s) - U_0(s) + Y_0(s),$$

$$\text{where} \quad \mathcal{Y}_0(s) = [\mathcal{Y}_1(s), \ldots, \mathcal{Y}_N(s)]^T,$$

$$\text{with} \quad \mathcal{Y}_i(s) = \sum_{j=1}^{N} \frac{a_{ij}}{d_i} \left(e^{-T_{ij}s} \int_{-T_{ij}}^{0} y_i(\eta) e^{-s\eta} d\eta - e^{-\tau_{ij}s} \int_{-\tau_{ij}}^{0} y_j(\eta) e^{-s\eta} d\eta \right)$$

and $Y(s) = [Y_1(s), \ldots, Y_N(s)]^T$. The initial conditions $Y_0(s) = [Y_{0,1}(s), \ldots, Y_{0,N}(s)]^T$ and $U_0(s) = [U_{0,1}(s), \ldots, U_{0,N}(s)]^T$ are defined above. Note that $\mathcal{Y}_0(s)$ is the initial condition resulting from the delays in the network for nonzero initial conditions. Reorganizing the above equation, we obtain

$$(\delta(s)I + \nu(s)\Gamma_r(s))Y(s) = -\nu(s)\mathcal{Y}_0(s) - U_0(s) + Y_0(s), \quad r = 1, 2, 3, 4, \tag{3.9}$$

where we have the initial conditions on the right and the solution $Y(s)$ on the left.

Using (3.9), we can now investigate if consensus solutions exist that satisfy $y_i(t) = y_j(t)$ for all i, j and all $t \in [-\mathcal{T}, \infty)$. This equality holds also for negative t because we consider identical initial conditions $y_i(\eta) = y_j(\eta)$ for all i, j and all $\eta \in [-\mathcal{T}, 0]$. The assumption of identical initial conditions is justified because for non-identical initial conditions with solutions satisfying $y_i(t) = y_j(t)$ for all i, j and all $t \geq 0$ we can shift our analysis to the starting point $t = \mathcal{T}$. Then, we have identical initial conditions $y_i(\eta) = y_j(\eta)$ for all i, j and all $\eta \in [0, \mathcal{T}]$.

If the initial conditions and solutions of all agents are identical, then we have $y_i(t) \equiv y_j(t) \Leftrightarrow Y_i(s) \equiv Y_j(s)$. Thus, we can reduce the analysis to just one agent, say agent i. Then, we have from (3.9)

$$\left(\delta(s) + \nu(s)\sum_{j=1}^{N}\frac{a_{ij}}{d_i}\left(e^{-T_{ij}s} - e^{-\tau_{ij}s}\right)\right)Y_i(s) = Y_0(s) - U_0(s)$$
$$- \nu(s)\sum_{j=1}^{N}\frac{a_{ij}}{d_i}\left(e^{-T_{ij}s}\int_{-T_{ij}}^{0} y_i(\eta)e^{-s\eta}d\eta - e^{-\tau_{ij}s}\int_{-\tau_{ij}}^{0} y_i(\eta)e^{-s\eta}d\eta\right),$$

for all $i \in \mathcal{N}$. Remember that $\delta(s), \nu(s), Y_0(s), U_0(s)$ are polynomials of finite degree. Now, we consider solutions $Y_i(s)$ that are rational functions in s, i.e. consensus solutions $y_i(t)$ that are a combination of polynomial, sinusoidal, and exponential functions in t. It is reasonable to focus on these functions because we are interested in consensus solutions that are solutions of the MAS for arbitrary delays $\tau_{ij}, T_{ij} \leq \mathcal{T}$, i.e. also for MAS without delays $\tau_{ij} = T_{ij} = 0$. In the latter case, we have $\delta(s)Y_i(s) = Y_0(s) - U_0(s)$, i.e. $Y_i(s)$ has to be a rational function. For rational $Y_i(s)$, we can separate the above equation in a purely rational part and a part with exponential terms

$$\delta(s)Y_i(s) = Y_0(s) - U_0(s) \tag{3.10}$$

$$0 = \nu(s)\sum_{j=1}^{N}\frac{a_{ij}}{d_i}\left(e^{-T_{ij}s}\left(Y_i(s) + \int_{-T_{ij}}^{0} y_i(\eta)e^{-s\eta}d\eta\right)\right.$$
$$\left.-e^{-\tau_{ij}s}\left(Y_i(s) + \int_{-\tau_{ij}}^{0} y_i(\eta)e^{-s\eta}d\eta\right)\right). \tag{3.11}$$

The first equation yields $Y_i(s) = \frac{Y_0(s) - U_0(s)}{\delta(s)}$, i.e. the numerator polynomial of $Y_i(s)$ is determined by the initial conditions and the denominator polynomial by the characteristic equation of the agent dynamics. By Assumption 3.1, all nonzero roots of $\delta(s)$ are in the open left half plane $\mathbb{C}^-$. Therefore, the inverse Laplace transform $y_i(t) = \mathcal{L}^{-1}(Y_i(s))$ satisfies $\lim_{t\to\infty} \dot{y}_i(t) \neq 0$ only if $\delta(s)$ has a root at least with multiplicity two at the origin.

3.3.2 Consensus Solutions

Now, we can analyse if MAS (3.1), (3.3) has consensus and flocking solutions. We start with consensus solutions where $y_i(t) = c$ for all i, all $t \geq -\mathcal{T}$, and any $c \in \mathbb{R}$. The Laplace transform is $Y_i(s) = \frac{c}{s}$ and we solve the integrals

$$\int_{-\tau_{ij}}^{0} y_i(\eta)e^{-s\eta}d\eta = \frac{c}{s}(e^{\tau_{ij}s} - 1) \qquad \int_{-T_{ij}}^{0} y_i(\eta)e^{-s\eta}d\eta = \frac{c}{s}(e^{T_{ij}s} - 1).$$

Thus, we obtain for the right hand side of (3.11)

$$\nu(s)\sum_{j=1}^{N}\frac{a_{ij}}{d_i}\left(e^{-T_{ij}s}\left(\frac{c}{s}+\frac{c}{s}(e^{T_{ij}s}-1)\right) - e^{-\tau_{ij}s}\left(\frac{c}{s}+\frac{c}{s}(e^{\tau_{ij}s}-1)\right)\right) = 0,$$

i.e. (3.11) holds for all $i \in \mathcal{N}$. Note moreover that $U_0(s) = 0$ if the initial conditions are constant, i.e. $y_i(t) = c$ for all i and all $t \in [-\mathcal{T}, 0]$, and

$$Y_0(s) = \sum_{l=1}^{n} \delta_l s^{l-1} c = \delta(s)\frac{c}{s} = \delta(s)Y_i(s),$$

because $\delta_0 = 0$ by Assumption 3.1. Therefore, consensus solutions result from any constant, identical initial condition.

In summary, MAS without self-delay, with identical self-delay, and with different self-delay have consensus solutions $y_i(t) = y_j(t) = \text{const.}$ for all i, j and all t. The attractivity of these solutions will be discussed in Section 3.4. Note that $u_i(t) = 0$ for consensus solutions and for all three feedback interconnections (3.3). In other words, the agents run in open loop and maintain identical outputs for identical initial conditions.

3.3.3 Flocking Solutions

Next, we investigate if these MAS also have flocking solutions, i.e. solutions $y_i(t) = y_j(t)$, for all i, j, t, with $\lim_{t\to\infty} \dot{y}_i(t) \neq 0$. As before, we consider rational Laplace transforms $Y_i(s) = Y_j(s)$ for all i, j, s. Remember that $Y_i(s) = \frac{Y_0(s)-U_0(s)}{\delta(s)}$, i.e. the poles of $Y_i(s)$ are the roots of $\delta(s)$. Since all nonzero roots of $\delta(s)$ are in $\mathbb{C}^-$, the MAS has flocking solutions only if $\delta(s)$ has multiple roots at the origin. Consider the simplest case $\delta(s) = s^2$. Thus, we have $Y_i(s) = \frac{Y_0(s)-U_0(s)}{s^2}$ and $y_i(t) = \mathcal{L}^{-1}(Y_i(s)) = c_1 t + c_0$ for appropriately chosen c_1, c_0. Solving again the integrals in $\mathcal{Y}_i(s)$ leads to

$$\int_{-\tau_{ij}}^{0} y_i(\eta)e^{-s\eta}d\eta = \frac{c_0}{s}(e^{\tau_{ij}s} - 1) + \frac{c_1}{s^2}\left(e^{\tau_{ij}s}(1+\tau_{ij}s) - 1\right) \tag{3.12a}$$

$$\int_{-T_{ij}}^{0} y_i(\eta)e^{-s\eta}d\eta = \frac{c_0}{s}(e^{T_{ij}s} - 1) + \frac{c_1}{s^2}\left(e^{T_{ij}s}(1+T_{ij}s) - 1\right). \tag{3.12b}$$

With this, we compute the right hand side of (3.11) and obtain

$$\nu(s)\sum_{j=1}^{N}\frac{a_{ij}}{d_i}\frac{c_1}{s^2}\left((T_{ij}-\tau_{ij})s-e^{-T_{ij}s}+e^{-\tau_{ij}s}\right),$$

which is in general nonzero for some $s \in \mathbb{C}$ and some $i \in \mathcal{N}$. This implies that (3.11) is not satisfied for flocking solutions in general, i.e. the MAS does not have flocking solutions. There is however a prominent case where (3.11) holds: for $\tau_{ij} = T_{ij}$, i.e. for identical self-delay. In this case, it is straightforward to compute c_1, c_0 from (3.10) based on the initial conditions using partial fractional decomposition.

If we consider more complicated agent dynamics where $\delta(s)$ has multiple roots at the origin, we determine $y_i(t) = \mathcal{L}^{-1}(Y_i(s))$ from (3.10) using partial fractional decomposition. In this case, $y_i(t)$ always contains polynomials in t, i.e. $y_i(t) = c_0 + c_1 t + c_2 t^2 + \ldots$, and additional exponentially decaying terms in t if $\delta(s)$ has roots in $\mathbb{C}^-$. Solving the integrals in $\mathcal{Y}_i(s)$ will always lead to summands as in (3.12). Therefore, the right hand side of (3.11) is in general nonzero for some $s \in \mathbb{C}$ and some $i \in \mathcal{N}$. Hence, (3.11) is not satisfied for flocking solutions except for the case of identical self-delay.

It turns out that flocking is in general only possible for MAS with identical self-delay but not for MAS without self-delay or MAS with different self-delay. It is interesting to know that flocking solutions $y_i(t) = y_j(t)$, for all i, j, t also lead to $u_i(t) = 0$, for all i, t for MAS with identical self-delay (3.3b) even if $\dot{y}_i(t) \neq 0$. Yet, we have in general $u_i(t) \neq 0$ for some i, t if $\dot{y}_i(t) \neq 0$ in MAS without self-delay (3.3a) or with different self-delay (3.3c).

3.3.4 Average Consensus and Average Flocking

MAS with identical self-delay have another interesting property: the average output is a solution of the agent dynamics H with zero input and average initial condition if the graph is regular, i.e. $d_i = d_j$, and the delays are symmetric, i.e. $\tau_{ij} = \tau_{ji}$. This is explained in the sequel.

Consider the average output of all MAS $\overline{y}(t) = \frac{1}{N}\sum_{i=1}^{N} y_i(t)$. The corresponding Laplace transform is $\overline{Y}(s) = \frac{1}{N}\sum_{i=1}^{N} Y_i(s) = \frac{1}{N}\mathbf{1}^T Y(s)$ where $\mathbf{1}^T = [1, 1, \ldots, 1]$. Now, we multiply $\mathbf{1}^T$ to the left of (3.9) and obtain

$$\mathbf{1}^T(\delta(s)+\nu(s)\Gamma_4(s))Y(s) = \mathbf{1}^T(Y_0(s)-\nu(s)\mathcal{Y}_0(s)-U_0(s)). \tag{3.13}$$

If the delays are symmetric and the graph is balanced, then $\Gamma_4(s)$ is symmetric for all $s \in \mathbb{C}$ and $\mathbf{1}^T$ is also its left eigenvector to the eigenvalue 0 because $\mathbf{1}^T\Gamma_4(s) = \mathbf{1}^T\Gamma_4^T(s) = (\Gamma_4(s)\mathbf{1})^T = 0$. Thus, we have

$$\mathbf{1}^T(\delta(s)+\nu(s)\Gamma_4(s))Y(s) = \mathbf{1}^T\delta(s)Y(s).$$

Moreover, as $a_{ij} = a_{ji}, d_i = d_j, \tau_{ij} = \tau_{ji}$, we have

$$\mathbf{1}^T \mathcal{Y}_0(s) = \sum_{i=1}^{N}\sum_{j=1}^{N} \frac{a_{ij}}{d_i}\left(e^{-\tau_{ij}s}\int_{-\tau_{ij}}^{0} y_i(\eta)e^{-s\eta}d\eta - e^{-\tau_{ij}s}\int_{-\tau_{ij}}^{0} y_j(\eta)e^{-s\eta}d\eta\right)$$

$$= \sum_{i=1}^{N}\sum_{j=1}^{N} \frac{a_{ij}}{d_i}\left(e^{-\tau_{ij}s}\int_{-\tau_{ij}}^{0} y_i(\eta)e^{-s\eta}d\eta - e^{-\tau_{ij}s}\int_{-\tau_{ij}}^{0} y_i(\eta)e^{-s\eta}d\eta\right) = 0,$$

$$\sum_{i=1}^{N} u_i^{(\iota)}(0) = -\sum_{i=1}^{N}\sum_{j=1}^{N} \frac{a_{ij}}{d_i}(y_i^{(\iota)}(-\tau_{ij}) - y_j^{(\iota)}(-\tau_{ij}))$$

$$= -\sum_{i=1}^{N}\sum_{j=1}^{N} \frac{a_{ij}}{d_i}(y_i^{(\iota)}(-\tau_{ij}) - y_i^{(\iota)}(-\tau_{ij})) = 0,$$

where $u_i^{(\iota)}$ is the ι-th derivative of u_i. We conclude $\mathbf{1}^T U_0(s) = 0$. Thus, Equation (3.13) simplifies to

$$\delta(s)\mathbf{1}^T Y(s) = \mathbf{1}^T Y_0(s),$$

i.e. the average output $\overline{y}(t)$ is a solution of the agent dynamics $H(s)$ for zero input and with average initial condition $\overline{y}^{(\iota)}(0), \iota = 0, 1, \dots, n$.

A similar result is obtained for MAS without self-delay and MAS with different self-delay if all delays are identical, i.e. $\tau_{ij} = \tau_{ji} = \tau$ and $T_{ij} = T_{ji} = T$, and the initial conditions are constant, i.e. $y_i(\eta) = y_i(0)$ for all $\eta \in [-\mathcal{T}, 0]$. As before, we assume that the graph is regular, i.e. $d_i = d_j$. Then, we multiply $\mathbf{1}^T$ to the left of (3.9) and obtain

$$\mathbf{1}^T(\delta(s) + \nu(s)\Gamma_3(s))Y(s) = \mathbf{1}^T(Y_0(s) - \nu(s)\mathcal{Y}_0(s) - U_0(s)). \tag{3.14}$$

Since the graph is regular and symmetric, the delays are identical, and the initial conditions are constant, we have

$$\nu(s)\mathbf{1}^T\Gamma_3(s)Y(s) = \nu(s)(e^{-Ts} - e^{-\tau s})\mathbf{1}^T Y(s),$$

$$\mathbf{1}^T Y_0(s) = \delta(s)\frac{1}{s}\mathbf{1}^T y(0),$$

$$\mathbf{1}^T \mathcal{Y}_0(s) = \sum_{i=1}^{N}\sum_{j=1}^{N} \frac{a_{ij}}{d_i}\left(\frac{y_i(0)}{s}(1 - e^{-Ts}) - \frac{y_j(0)}{s}(1 - e^{-\tau s})\right)$$

$$= \frac{1}{s}(1 - e^{-Ts})\mathbf{1}^T y(0) - (1 - e^{-\tau s})\mathbf{1}^T y(0) = \frac{1}{s}(e^{-\tau s} - e^{-Ts})\mathbf{1}^T y(0),$$

$$\sum_{i=1}^{N} u_i^{(\iota)}(0) = -\sum_{i=1}^{N}\sum_{j=1}^{N} \frac{a_{ij}}{d_i}(y_i^{(\iota)}(-T_{ij}) - y_j^{(\iota)}(-\tau_{ij})) = 0,$$

i.e. $\mathbf{1}^T U_0(s) = 0$. Thus, Equation (3.14) simplifies to

$$(\delta(s) + \nu(s)(e^{-Ts} - e^{-\tau s}))\mathbf{1}^T Y(s) = (\delta(s) + \nu(s)(e^{-Ts} - e^{-\tau s}))\frac{1}{s}\mathbf{1}^T y(0),$$

Table 3.1: Summary of consensus properties of MAS with identical dynamics H.

	MAS without self-delay (3.3a)	MAS with identical self-delay (3.3b)	MAS with different self-delay (3.3c)
consensus	yes	yes	yes
flocking	no	yes	no
average cons.	if $d_i = d_j, \tau_{ij} = \tau$	if $d_i = d_j, \tau_{ij} = \tau_{ji}$	if $d_i = d_j, \tau_{ij} = \tau, T_{ij} = T$

which is true if $\mathbf{1}^T Y(s) = \frac{1}{s}\mathbf{1}^T y(0)$. In other words, the average output is a constant, i.e. $\overline{y}(t) = \overline{y}(0)$.

Consider the special case of single integrator agents, i.e. $H(s) = \frac{K}{s}, K > 0$. Then, we have $\overline{y}(t) = \overline{y}(0)$ also for MAS with identical self-delays, regular graphs, and symmetric delays. That is, single integrator MAS with identical self-delay interconnected on regular graphs with symmetric delays achieve *average consensus*, i.e. $\lim_{t\to\infty} y_i(t) = \overline{y}(0) = \frac{1}{N}\sum_{j=1}^{N} y_j(0)$ for all i if suitable conditions on K and $\mathcal{T}$ hold, see Corollary 3.10 and Bliman and Ferrari-Trecate (2008). For MAS without self-delay, average consensus on regular graphs and with identical delays $\tau_{ij} = \tau_{ji} = \tau$ is achieved for arbitrary delays and gains K, see Corollary 3.9 and Seuret et al. (2008).

Both average consensus results are extensions of results in Bliman and Ferrari-Trecate (2008); Seuret et al. (2008) that deal with single integrator MAS. Our results also show that flocking is only possible if the agent dynamics H have multiple poles at the origin. Remember that H cannot have poles in the open right half plane due to Assumption 3.1.

The different consensus and flocking solutions of the different delay configurations (3.3) investigated in this section are summarized in Table 3.1. Moreover, we illustrate these findings in the following examples.

Example 3.3 (Flocking in MAS with and without self-delay). *First, we investigate the flocking of double integrator MAS of the following form*

$$\ddot{y}_i(t) = -K_1 \sum_{j=1}^{N} \frac{a_{ij}}{d_i}(\dot{y}_i(t) - \dot{y}_j(t-\tau_{ij})) - K_0 \sum_{j=1}^{N} \frac{a_{ij}}{d_i}(y_i(t) - y_j(t-\tau_{ij})), \tag{3.15a}$$

$$\ddot{y}_i(t) = -K_1 \sum_{j=1}^{N} \frac{a_{ij}}{d_i}(\dot{y}_i(t-\tau_{ij}) - \dot{y}_j(t-\tau_{ij})) - K_0 \sum_{j=1}^{N} \frac{a_{ij}}{d_i}(y_i(t-\tau_{ij}) - y_j(t-\tau_{ij})), \tag{3.15b}$$

i.e. with identical agent dynamics $H(s) = \frac{K_1 s + K_0}{s^2}, K_1 > 0, K_0 > 0$. *We consider here only MAS without self-delay* (3.15a) *and with identical self-delay* (3.15b). *Similar results can be obtained for MAS with different self-delay.*

In this example, we are not interested in conditions on K_0, K_1, *and* $\mathcal{T}$ *such that consensus or flocking is achieved. We chose* $K_0 = K_1 = 0.7$ *and* $\mathcal{T} = 0.5$ *such that the trajectories of the agents converge toward each other and discuss the consensus or flocking solutions. We simulate two MAS consisting of* $N = 10$ *agents on a ring topology with randomly generated delays* $\tau_{ij} \leq \mathcal{T} = 0.5$ *and constant initial conditions. Figure 3.1(a)*

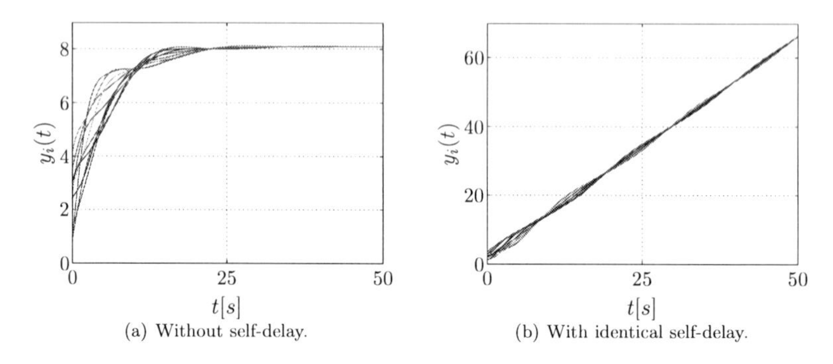

(a) Without self-delay.

(b) With identical self-delay.

Figure 3.1: Simulation results for MAS without self-delay and with identical self-delay with $N = 10$ agents, dynamics $H(s) = \frac{0.7s+0.7}{s^2}$, ring topology, and random delays $\tau_{ij} \leq \mathcal{T} = 0.5$, see Example 3.3.

shows the simulation results of the MAS without self-delay. Clearly, this MAS reaches consensus whereas the MAS with identical self-delay reaches flocking, see Figure 3.1(b), just as derived in this section. The remaining oscillations around the flocking solution in Figure 3.1(b) vanish over time.

Example 3.4 (Average consensus in MAS with and without self-delay). *Now, we study average consensus of single integrator MAS of the following form*

$$\dot{y}_i(t) = -K \sum_{j=1}^{N} \frac{a_{ij}}{d_i}(y_i(t) - y_j(t - \tau_{ij})), \tag{3.16}$$

$$\dot{y}_i(t) = -K \sum_{j=1}^{N} \frac{a_{ij}}{d_i}(y_i(t - \tau_{ij}) - y_j(t - \tau_{ij})), \tag{3.17}$$

$$\dot{y}_i(t) = -K \sum_{j=1}^{N} \frac{a_{ij}}{d_i}(y_i(t - T_{ij}) - y_j(t - \tau_{ij})), \tag{3.18}$$

i.e. with identical agent dynamics $H(s) = \frac{K}{s}, K > 0$. Conditions on K and $\mathcal{T}$ such that consensus is achieved will be derived in Section 3.5. Here, we choose $K = 0.7$ and $\mathcal{T} = 0.5$ such that the agents reach consensus. We simulate six MAS, two without self-delay, two with identical self-delay, and two with different self-delay. All MAS consist of $N = 10$ agents on a ring topology with randomly generated delays $\tau_{ij} \leq \mathcal{T} = 0.5$. Note that the ring topology implies that $d_i = d_j$ in all situations. Figure 3.2 shows the average output $\overline{y}(t) = \frac{1}{N}\sum_{i=1}^{N} y_i(t)$ of all agents.

In Figure 3.2(a), the average output of MAS without self-delay is displayed. Thereby, the dashed line is the average output of a MAS with heterogeneous, symmetric delays

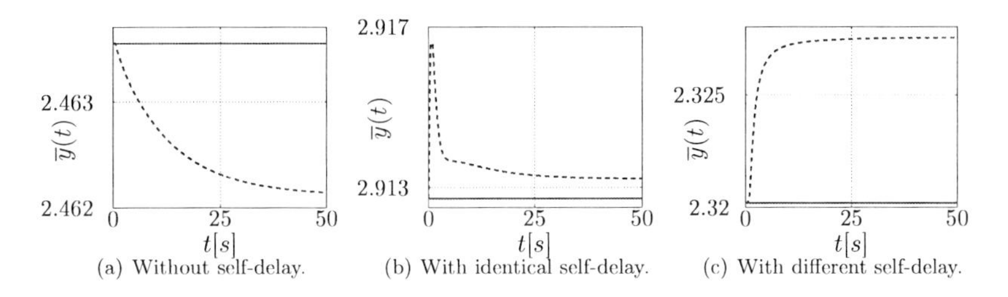

(a) Without self-delay. (b) With identical self-delay. (c) With different self-delay.

Figure 3.2: Average output $\overline{y}(t)$ for MAS with and without self-delay with $N = 10$ agents, dynamics $H(s) = \frac{0.7}{s}$, ring topology, and random delays $\tau_{ij}, T_{ij} \leq \mathcal{T} = 0.5$, see Example 3.4 for more details on the solid and dashed lines.

$\tau_{ij} = \tau_{ji} \leq \mathcal{T}$ and the solid line describes $\overline{y}(t)$ for a MAS with homogeneous delays $\tau_{ij} = \tau \leq \mathcal{T}$. We see that average consensus is only achieved for homogeneous delays, see also Table 3.1.

In Figure 3.2(b), the average output of MAS with identical self-delay is shown. The dashed line describes a MAS with asymmetric delays $\tau_{ij} \neq \tau_{ji}$ whereas the solid line results from a MAS with symmetric delays $\tau_{ij} = \tau_{ji} \leq \mathcal{T}$. Again, the findings of this section are confirmed, see Table 3.1. Average consensus is only obtained for symmetric delays.

Finally, Figure 3.2(c) illustrates the average output of MAS with different self-delay. The dashed line results from a MAS with heterogeneous, symmetric delays $\tau_{ij} = \tau_{ji} \leq \mathcal{T}, T_{ij} = T_{ji} \leq \mathcal{T}$; whereas the solid line comes from a MAS with homogeneous delays $\tau_{ij} = \tau \leq \mathcal{T}$ $T_{ij} = T \leq \mathcal{T}$. Once more, the results of this section are reflected in the simulations. Average consensus is achieved for homogeneous delays but not for heterogeneous delays.

Table 3.1 and Examples 3.3 and 3.4 nicely summarize the findings of this section. It is important to remember that the different delay configurations enable different cooperative control tasks, depending on the symmetry of the topology and the delays.

3.4 Generalized Nyquist Consensus Condition

3.4.1 Generalized Nyquist Criterion for Multi-Agent Systems

In the previous section, we investigated the existence of different consensus and flocking solutions of the MAS (3.1), (3.3) for appropriately chosen initial conditions. There, we considered identical agent dynamics. Straightforward calculations lead to the same results for non-identical agent dynamics. Remember that the following equation relates the solutions $Y(s)$ to the initial conditions $Y_0(s), U_0(s), \mathcal{Y}_0(s)$

$$Y(s) = (\operatorname{diag}(\delta_i(s)) + \operatorname{diag}(\nu_i(s))\Gamma_r(s))^{-1}(Y_0(s) - U_0(s) - \operatorname{diag}(\nu_i(s))\mathcal{Y}_0(s)), \quad (3.19)$$

for $r = 1, 2, 3, 4$, see (3.9), where $\delta(s)$ and $\nu(s)$ are replaced by $\text{diag}(\delta_i(s))$ and $\text{diag}(\nu_i(s))$ because we consider here the general case of non-identical agent dynamics $H_i(s)$. The more challenging task is now to provide conditions such that these consensus and flocking solutions are asymptotically attracting for all initial conditions. Therefore, we have to analyse the characteristic quasi-polynomial of the closed loop system (3.19)

$$\Delta(s) = \det\left(\text{diag}(\delta_i(s)) + \text{diag}(\nu_i(s))\Gamma_r(s)\right), \quad r = 1, 2, 3, 4. \tag{3.20}$$

The characteristic quasi-polynomial $\Delta(s)$ has a root at the origin, possibly with multiplicity larger then one, because $\delta_i(0) = 0$ for all i, by Assumption 3.1, and $\Gamma_r(0) = \overline{L} = I - D^{-1}A$ has an eigenvalue at the origin, see Appendix B. If the graph is connected, all roots at the origin correspond to the consensus and flocking solutions, as explained next.

The transfer function $(\text{diag}(\delta_i(s)) + \text{diag}(\nu_i(s))\Gamma_r(s))^{-1}$ in (3.19) is a matrix transfer function. Thus, any pole of this transfer function has a corresponding *input* and *output pole direction.* They describe in which input and output direction the pole affects the input-output behavior, see for example Skogestad and Postlethwaite (2004). In the present case, we are interested in the output pole direction of the pole of $(\text{diag}(\delta_i(s)) + \text{diag}(\nu_i(s))\Gamma_r(s))^{-1}$ at the origin, i.e. how that pole relates to the solutions $Y(s)$ of the MAS. This output pole direction corresponds to the input zero direction of $\text{diag}(\delta_i(s)) + \text{diag}(\nu_i(s))\Gamma_r(s)$, compare Skogestad and Postlethwaite (2004), i.e. the vector $Y_p \in \mathbb{R}^N$ such that

$$(\text{diag}(\delta_i(0)) + \text{diag}(\nu_i(0))\Gamma_r(0))Y_p = \text{diag}(\nu_i(0))\overline{L}Y_p = 0.$$

Note that $\nu_i(0) \neq 0$ because ν_i and δ_i are coprime and $\delta_i(0) = 0$. Therefore, the above equation is only true if Y_p is eigenvector of $\overline{L}$ with eigenvalue 0. If $\overline{L}$ is the Laplacian matrix of a connected graph, this is equivalent to $Y_p = c\mathbf{1}, c \in \mathbb{R}$. Remember that $Y(s) = Y_i(s)\mathbf{1}$ describes the consensus and flocking solutions, see Section 3.3. Therefore, all zero roots of $\Delta(s)$ correspond to the consensus and flocking solutions.

The consensus and flocking solutions are asymptotically attracting if all nonzero roots of $\Delta(s)$ are in $\mathbb{C}^-$. The computation of the roots of (3.20) is extremely difficult because $\Gamma_r(s)$ depends on the frequency dependent matrices $A_\tau(s), D_\tau(s), D_T(s)$. Instead of computing these roots, we use the *generalized Nyquist criterion* for distributed, i.e. non-rational, transfer functions, as proposed in Desoer and Wang (1980), to determine the asymptotic stability of the consensus solutions. The generalized Nyquist criterion relates the roots of $\Delta(s) = \det(I + G_r(s))$ in $\mathbb{C}^+$ to the number of clockwise encirclements of -1 by the eigenloci of the open loop transfer function $G_r(s)$, see Figure 3.3. Here, we denote G_r *network return ratio.* We propose the following network return ratio

$$G_r(s) = -\text{diag}\left(\frac{\nu_i(s)}{\delta_i(s) + 2\nu_i(s)}\right)(2I - \Gamma_r(s)), \quad r = 1, 2, 3, 4. \tag{3.21}$$

We recover the closed loop system (3.19) with inputs and outputs $\tilde{Y}(s)$ and $\tilde{U}(s)$, where

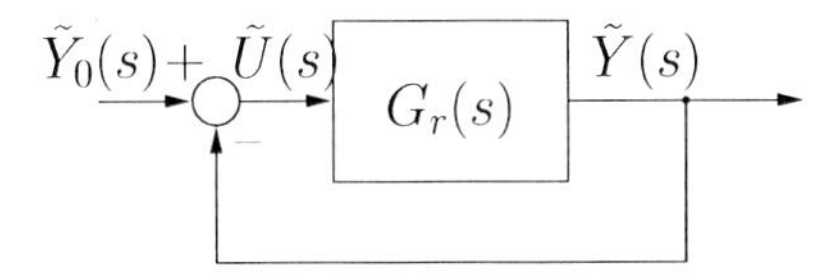

Figure 3.3: General negative feedback interconnection.

$\tilde{U}(s) = \tilde{Y}_0(s) - \tilde{Y}(s)$, as follows

$$\begin{aligned}
\operatorname{diag}(\delta_i(s) + 2\nu_i(s))\tilde{Y}(s) &= -\operatorname{diag}(\nu_i(s))(2I - \Gamma_r(s))\tilde{U}(s) \\
&= -\operatorname{diag}(\nu_i(s))(2I - \Gamma_r(s))(\tilde{Y}_0(s) - \tilde{Y}(s)) \\
(\operatorname{diag}(\delta_i(s)) + \operatorname{diag}(\nu_i(s))\Gamma_r(s))\tilde{Y}(s) &= -\operatorname{diag}(\nu_i(s))(2I - \Gamma_r(s))\tilde{Y}_0(s), \quad r = 1,2,3,4,
\end{aligned}$$

where $\tilde{Y}(s) = Y(s)$ and $\operatorname{diag}(\nu_i(s))(2I - \Gamma_r(s))\tilde{Y}_0(s) = U_0(s) + \operatorname{diag}(\nu_i(s))\mathcal{Y}_0(s) - Y_0(s)$. Note that the return ratio G_r is asymptotically stable due to Assumption 3.1. This is an important advantage in the sequel, in particular compared to alternative return ratios like $G_r(s) = \operatorname{diag}(H_i(s))\Gamma_r(s)$ that has a pole at the origin. In order to apply the generalized Nyquist criterion, we have to compute the eigenloci of $G_r(j\omega)$. We will see later on that the exact eigenloci need not be determined if we have appropriate sets containing the eigenvalues of $2I - \Gamma_r(j\omega)$. These sets are determined in the next subsection.

3.4.2 Spectrum of $2I - \Gamma_r(j\omega)$

Now, we provide convex sets $\Omega_r(\omega\mathcal{T})$ that contain the eigenvalues of $2I - \Gamma_r(j\omega)$ for arbitrary, undirected topologies and arbitrary, bounded delays $\tau_{ij}, T_{ij} \leq \mathcal{T}$. These sets are described using convex hulls, see for example Boyd and Vandenberghe (2004). The *convex hull* of a set $\{\zeta(\chi) : \chi \in \Delta\chi\}$, where $\Delta\chi = [\underline{\chi}_1, \overline{\chi}_1] \times \ldots \times [\underline{\chi}_{n_\chi}, \overline{\chi}_{n_\chi}] \subset \mathbb{R}^{n_\chi}$, is

$$\operatorname{Co}\{\zeta(\chi), \chi \in \Delta\chi\} = \left\{ \int_{\Delta\chi} \phi(\chi)\zeta(\chi)d\chi : \chi \in \Delta\chi, \phi(\chi) \geq 0, \int_{\Delta\chi} \phi(\chi)d\chi = 1 \right\}. \quad (3.22)$$

The sets Ω_r are defined as follows

$$\Omega_1(\omega\mathcal{T}) = \operatorname{Co}\left\{1 - e^{-j\chi_1}, 1 + e^{-j\chi_2} : \chi_1, \chi_2 \in [0, \omega\mathcal{T}]\right\}, \quad (3.23a)$$

$$\Omega_2(\omega\mathcal{T}) = \operatorname{Co}\left\{2 - e^{-j\chi_1} + e^{-j\frac{\chi_1+\chi_2}{2}}, 2 - e^{-j\chi_3} - e^{-j\frac{\chi_3+\chi_4}{2}} : \chi_1, \chi_2, \chi_3, \chi_4 \in [0, \omega\mathcal{T}]\right\}, \quad (3.23b)$$

$$\Omega_3(\omega\mathcal{T}) = \operatorname{Co}\left\{2 - e^{-j\chi_1} + e^{-j\chi_2}, 2 - e^{-j\chi_3} - e^{-j\chi_4} : \chi_1, \chi_2, \chi_3, \chi_4 \in [0, \omega\mathcal{T}]\right\}, \quad (3.23c)$$

$$\Omega_4(\omega\mathcal{T}) = \operatorname{Co}\left\{2 - 2e^{-j\chi}, 2 : \chi \in [0, \omega\mathcal{T}]\right\}. \quad (3.23d)$$

These sets are illustrated in Figure 3.4 and 3.5. With these sets, we have the following lemma:

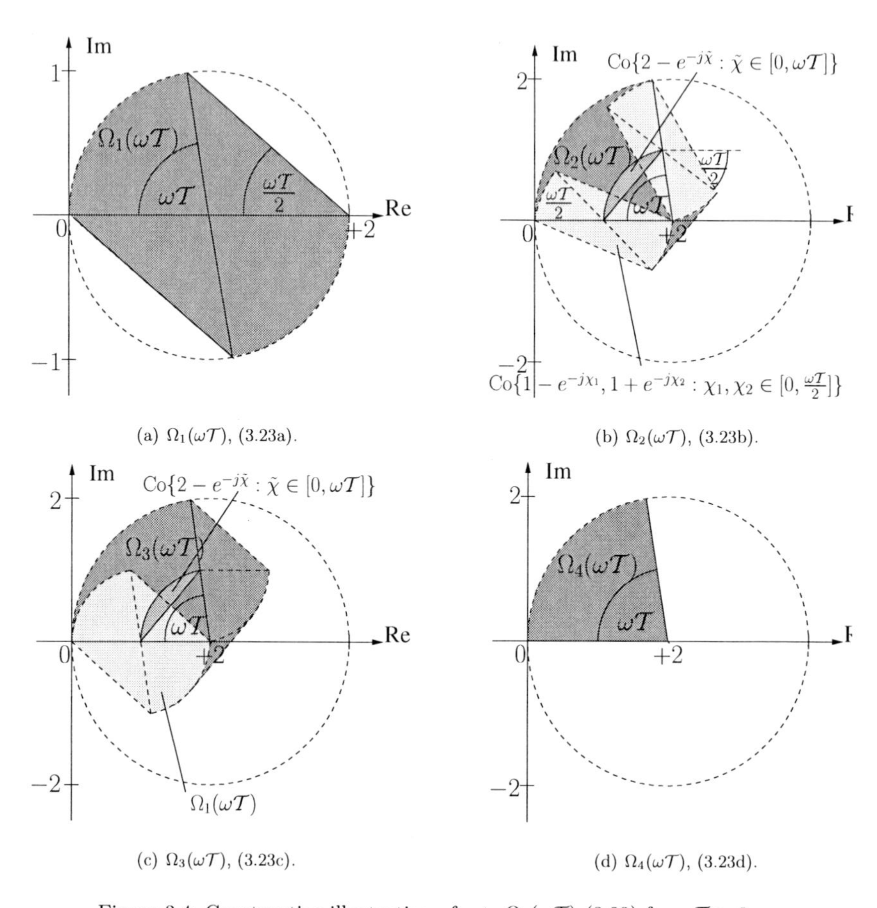

Figure 3.4: Constructive illustration of sets $\Omega_r(\omega\mathcal{T})$ (3.23) for $\omega\mathcal{T} > 0$.

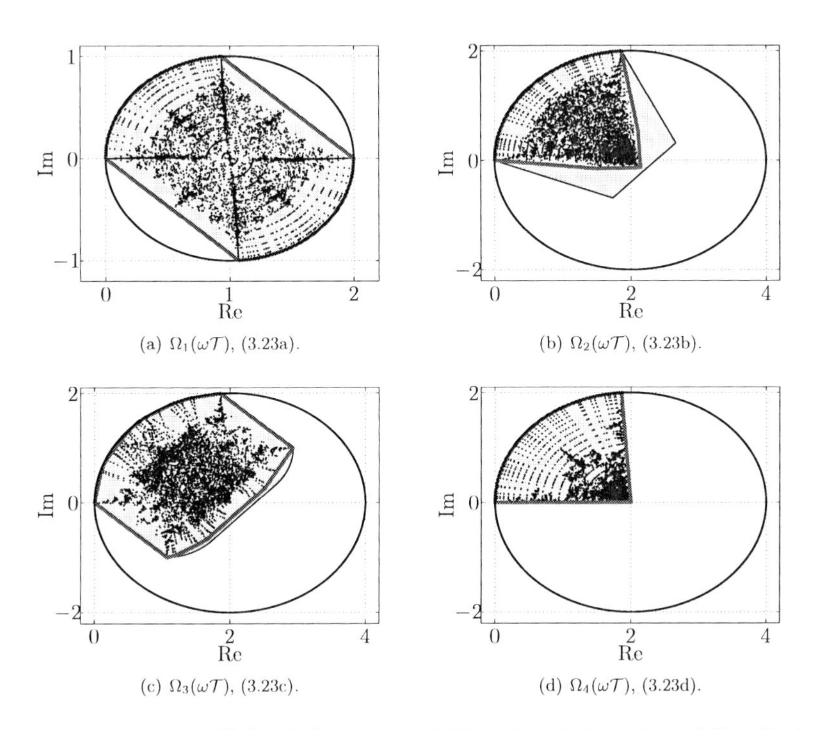

(a) $\Omega_1(\omega\mathcal{T})$, (3.23a).

(b) $\Omega_2(\omega\mathcal{T})$, (3.23b).

(c) $\Omega_3(\omega\mathcal{T})$, (3.23c).

(d) $\Omega_4(\omega\mathcal{T})$, (3.23d).

Figure 3.5: Sets $\Omega_r(\omega\mathcal{T})$ (3.23) for $\omega = 3$ and $\mathcal{T} = 0.5$ and eigenvalues of 50×50 ring topology matrix with 150 randomly chosen sets of delays.

Lemma 3.5 (Spectrum of $2I - \Gamma_r(j\omega)$). *For arbitrary bounded delays $\tau_{ij}, T_{ij} \leq \mathcal{T}$ and arbitrary undirected topologies, the spectrum $\sigma(2I - \Gamma_r(j\omega))$ satisfies for MAS*

$$\begin{aligned}
&\textit{without self-delay} && \sigma(2I - \Gamma_1(j\omega)) \subset \Omega_1(\omega\mathcal{T}), && (3.24)\\
&\textit{with identical self-delay} && \sigma(2I - \Gamma_2(j\omega)) \subset \Omega_2(\omega\mathcal{T}), && (3.25)\\
&\textit{with different self-delay} && \sigma(2I - \Gamma_3(j\omega)) \subset \Omega_3(\omega\mathcal{T}). && (3.26)
\end{aligned}$$

If the delays are symmetric, i.e. $\tau_{ij} = \tau_{ji} \leq \mathcal{T}$ and $T_{ij} = T_{ji} \leq \mathcal{T}$, then we have for MAS

$$\begin{aligned}
&\textit{without self-delay} && \sigma(2I - \Gamma_1(j\omega)) \subset \Omega_1(\omega\mathcal{T}), && (3.27)\\
&\textit{with identical self-delay} && \sigma(2I - \Gamma_4(j\omega)) \subset \Omega_4(\omega\mathcal{T}), && (3.28)\\
&\textit{with different self-delay} && \sigma(2I - \Gamma_3(j\omega)) \subset \Omega_3(\omega\mathcal{T}). && (3.29)
\end{aligned}$$

Proof. The proof uses the *field of values* of a matrix, see e.g. Horn and Johnson (1991). The field of values $\mathcal{F}$ of a matrix $M \in \mathbb{C}^{N\times N}$ is defined as

$$\mathcal{F}(M) = \{v^* M v : v \in \mathbb{C}^N, v^* v = 1\},$$

where v^* denotes the complex conjugate of v. The most important property of the field of values is that it contains the spectrum of M, i.e. $\sigma(M) \subset \mathcal{F}(M)$. Moreover, we will use the fact that $\sigma(D^{-1}M) = \sigma(D^{-\frac{1}{2}} M D^{-\frac{1}{2}})$ for any diagonal positive definite matrix $D \in \mathbb{R}^{N\times N}$ and any matrix $M \in \mathbb{C}^{N\times N}$. This results directly form the definition of the eigenvalues $\lambda = \lambda(D^{-1}M)$:

$$\lambda v = D^{-1} M v \quad \Leftrightarrow \quad \lambda D^{\frac{1}{2}} v = D^{-\frac{1}{2}} M v \quad \Leftrightarrow \quad \lambda \tilde{v} = D^{-\frac{1}{2}} M D^{-\frac{1}{2}} \tilde{v},$$

where $\tilde{v} = D^{\frac{1}{2}} v$. Note that the spectrum of $D^{-1}M$ is real if M is real symmetric because $D^{-\frac{1}{2}} M D^{-\frac{1}{2}}$ is real symmetric.

Spectrum of $2I - \Gamma_1(j\omega)$ First, we prove (3.24). From our above arguments, we have $\sigma(D^{-1}A_\tau(j\omega)) \subset \mathcal{F}\left(D^{-\frac{1}{2}} A_\tau(j\omega) D^{-\frac{1}{2}}\right)$. For any $v \in \mathbb{C}^N$ with $v^* v = 1$, we have

$$\begin{aligned}
v^* D^{-\frac{1}{2}} A_\tau(j\omega) D^{-\frac{1}{2}} v &= \sum_{i=1}^{N}\sum_{j=1}^{N} a_{ij} e^{-j\omega\tau_{ij}} \frac{v_i^*}{\sqrt{d_i}} \frac{v_j}{\sqrt{d_j}}\\
&= \frac{1}{2}\sum_{i=1}^{N}\sum_{j=1}^{N} a_{ij} e^{-j\omega\hat{\tau}_{ij}} \left(e^{-j\omega\check{\tau}_{ij}} \frac{v_i^* v_j}{\sqrt{d_i d_j}} + e^{j\omega\check{\tau}_{ij}} \frac{v_j^* v_i}{\sqrt{d_j d_i}} \right),
\end{aligned}$$

where $\hat{\tau}_{ij} = \frac{\tau_{ij}+\tau_{ji}}{2}$, $\check{\tau}_{ij} = \frac{\tau_{ij}-\tau_{ji}}{2}$, and we used the fact that $a_{ij} = a_{ji}$. Note that the term in parenthesis is the sum of a complex number and its complex conjugate, i.e. twice the real part of this complex number. The real part of a complex number z is $\mathrm{Re}\{z\} = \tilde{\xi}|z|$ for some $\tilde{\xi} \in [-1, 1]$. Thus, we have

$$2\mathrm{Re}\left\{ e^{-j\omega\check{\tau}_{ij}} \frac{v_i^* v_j}{\sqrt{d_i d_j}} \right\} = 2\tilde{\xi}_{ij} \frac{|v_i^* v_j|}{\sqrt{d_i d_j}} = \xi_{ij} \left(\frac{|v_i|^2}{d_i} + \frac{|v_j|^2}{d_j} \right),$$

for some $\tilde{\xi}_{ij} = \tilde{\xi}_{ji} \in [-1,1], \xi_{ij} = \xi_{ji} \in [-1,1]$, where we use the triangle inequality in the second step. We obtain

$$v^* D^{-\frac{1}{2}} A_\tau(j\omega) D^{-\frac{1}{2}} v = \frac{1}{2}\sum_{i=1}^{N}\sum_{j=1}^{N} a_{ij} e^{-j\omega\hat{\tau}_{ij}} \xi_{ij} \left(\frac{|v_i|^2}{d_i} + \frac{|v_j|^2}{d_j}\right)$$
$$= \sum_{i=1}^{N}\sum_{j=1}^{N} \frac{a_{ij}}{d_i} e^{-j\omega\hat{\tau}_{ij}} \xi_{ij} |v_i|^2.$$

Thus, the spectrum of $D^{-1}A_\tau(j\omega)$ satisfies

$$\sigma(D^{-1}A_\tau(j\omega)) \subset \mathcal{F}\left(D^{-\frac{1}{2}} A_\tau(j\omega) D^{-\frac{1}{2}}\right) \subseteq \mathrm{Co}\left\{-e^{-j\chi_1}, +e^{-j\chi_2} : \chi_1, \chi_2 \in [0, \omega\mathcal{T}]\right\},$$

because $\sum_{j=1}^{N} \frac{a_{ij}}{d_i} = 1, \sum_{i=1}^{N} |v_i|^2 = 1, \xi_{ij} = \xi_{ji} \in [-1,1]$, and $\hat{\tau}_{ij} \in [0, \mathcal{T}]$, for all i, j. Thus, the eigenvalues of $2I - \Gamma_1(j\omega) = I + D^{-1}A_\tau(j\omega)$ are in $\Omega_1(\omega\mathcal{T})$ because $\sigma(I + D^{-1}A_\tau) = 1 - \sigma(D^{-1}A_\tau)$.

Spectrum of $2I - \Gamma_2(j\omega)$ Next, we prove (3.25). By our former arguments, we have $\sigma(D^{-1}L_\tau(j\omega)) \subset \mathcal{F}(D^{-\frac{1}{2}} L_\tau(j\omega) D^{-\frac{1}{2}})$. For any $v \in \mathbb{C}^N$ with $v^* v = 1$, we have

$$v^* D^{-\frac{1}{2}} L_\tau(j\omega) D^{-\frac{1}{2}} v = \sum_{i=1}^{N}\sum_{j=1}^{N} a_{ij} e^{-j\omega\tau_{ij}} \left(\frac{|v_i|^2}{d_i} - \frac{v_i^*}{\sqrt{d_i}} \frac{v_j}{\sqrt{d_j}}\right)$$
$$= \sum_{i=1}^{N}\sum_{j=1}^{N} \frac{a_{ij}}{d_i} \left(e^{-j\omega\tau_{ij}} - \xi_{ij} e^{-j\omega\hat{\tau}_{ij}}\right) |v_i|^2,$$

where $\hat{\tau}_{ij} = \frac{\tau_{ij} + \tau_{ji}}{2}$ and $\xi_{ij} = \xi_{ji} \in [-1,1]$ as above. Hence, the spectrum of $D^{-1}L_\tau(j\omega)$ satisfies

$$\sigma(D^{-1}L_\tau(j\omega)) \subset \mathcal{F}\left(D^{-\frac{1}{2}} L_\tau(j\omega) D^{-\frac{1}{2}}\right)$$
$$\subseteq \mathrm{Co}\left\{e^{-j\chi_1} - e^{-j\frac{\chi_1+\chi_2}{2}}, e^{-j\chi_3} + e^{-j\frac{\chi_3+\chi_4}{2}} : \chi_1, \chi_2, \chi_3, \chi_4 \in [0, \omega\mathcal{T}]\right\},$$

because $\sum_{j=1}^{N} \frac{a_{ij}}{d_i} = 1, \sum_{i=1}^{N} |v_i|^2 = 1, \xi_{ij} = \xi_{ji} \in [-1,1], \tau_{ij} \in [0, \mathcal{T}]$ for all i, j. Thus, we have $\sigma(2I - \Gamma_2(j\omega)) = \sigma(2I - D^{-1}L_\tau(j\omega)) \subset \Omega_2(\omega\mathcal{T})$.

Spectrum of $2I - \Gamma_3(j\omega)$ Now, we prove (3.26). Combining the above results, we have

$$\sigma(D^{-1}L_{\tau T}(j\omega)) \subset \mathcal{F}\left(D^{-\frac{1}{2}} D_T(j\omega) D^{-\frac{1}{2}}\right) + \mathcal{F}\left(D^{-\frac{1}{2}} A_\tau(j\omega) D^{-\frac{1}{2}}\right), \tag{3.30}$$

because $\mathcal{F}(M_1 + M_2) \subseteq \mathcal{F}(M_1) + \mathcal{F}(M_2)$, where $+$ indicates an element-wise sum of the two sets. Note moreover that

$$v^* D^{-\frac{1}{2}} D_T(j\omega) D^{-\frac{1}{2}} v = \sum_{i=1}^{N}\sum_{j=1}^{N} \frac{a_{ij}}{d_i} e^{-j\omega T_{ij}} |v_i|^2,$$

i.e. $\mathcal{F}\left(D^{-\frac{1}{2}}D_{\mathcal{T}}(j\omega)D^{-\frac{1}{2}}\right) = \text{Co}\left\{e^{-j\chi} : \chi \in [0, \omega\mathcal{T}]\right\}$. Together with our previous results for $\mathcal{F}\left(D^{-\frac{1}{2}}A_{\tau}(j\omega)D^{-\frac{1}{2}}\right)$, we obtain $\mathcal{F}(2I - D^{-\frac{1}{2}}L_{\tau T}(j\omega)D^{-\frac{1}{2}} \subseteq \Omega_3(\omega\mathcal{T})$.

Spectrum of $2I - \Gamma_4(j\omega)$ Finally, we prove (3.28). We know that $\sigma(D^{-1}L_{\tau}(s)) \in \mathcal{F}(D^{-\frac{1}{2}}L_{\tau}(s)D^{-\frac{1}{2}})$. For any $v \in \mathbb{C}^N$ with $v^*v = 1$, we have

$$
\begin{aligned}
v^*D^{-\frac{1}{2}}L_{\tau}(j\omega)D^{-\frac{1}{2}}v &= \sum_{i=1}^{N}\sum_{j=1}^{N} a_{ij}e^{-j\omega\tau_{ij}}\frac{|v_i|^2}{d_i} - \sum_{i=1}^{N}\sum_{j=1}^{N} a_{ij}e^{-j\omega\tau_{ij}}\frac{v_i^*}{\sqrt{d_i}}\frac{v_j}{\sqrt{d_j}} \\
&= \frac{1}{2}\sum_{i=1}^{N}\sum_{j=1}^{N} a_{ij}e^{-j\omega\tau_{ij}}\left(\frac{|v_i|^2}{d_i} - \frac{v_i^*}{\sqrt{d_i}}\frac{v_j}{\sqrt{d_j}} - \frac{v_j^*}{\sqrt{d_j}}\frac{v_i}{\sqrt{d_i}} + \frac{|v_j|^2}{d_j}\right)
\end{aligned}
$$

where we used the fact that $a_{ij} = a_{ji}$ and $\tau_{ij} = \tau_{ji}$. The two elements in parenthesis $\frac{v_i^*}{\sqrt{d_i}}\frac{v_j}{\sqrt{d_j}}$ and $\frac{v_j^*}{\sqrt{d_j}}\frac{v_i}{\sqrt{d_i}}$ form again the sum of a complex number and its conjugate. We obtain

$$
-2\text{Re}\left\{\frac{v_i^*v_j}{\sqrt{d_id_j}}\right\} = 2\tilde{\xi}_{ij}\frac{|v_i^*v_j|}{\sqrt{d_id_j}} = \xi_{ij}\left(\frac{|v_i|^2}{d_i} + \frac{|v_j|^2}{d_j}\right)
$$

for some $\tilde{\xi}_{ij} = \tilde{\xi}_{ji} \in [-1, 1], \xi_{ij} = \xi_{ji} \in [-1, 1]$. Thus, we have

$$
\begin{aligned}
v^*D^{-\frac{1}{2}}L_{\tau}(j\omega)D^{-\frac{1}{2}}v &= \frac{1}{2}\sum_{i=1}^{N}\sum_{j=1}^{N} a_{ij}e^{-j\omega\tau_{ij}}(1 + \xi_{ij})\left(\frac{|v_i|^2}{d_i} + \frac{|v_j|^2}{d_j}\right) \\
&= \sum_{i=1}^{N}\sum_{j=1}^{N}\frac{a_{ij}}{d_i}e^{-j\omega\tau_{ij}}(1 + \xi_{ij})|v_i|^2,
\end{aligned}
$$

and we obtain

$$
\sigma(D^{-1}L_{\tau}(j\omega)) \subset \text{Co}\left\{2e^{-j\chi}, 0 : \chi \in [0, \omega\mathcal{T}]\right\}.
$$

In summary, we have $\sigma(2I - D^{-1}L_{\tau}(j\omega)) \subset \Omega_4(\omega\mathcal{T})$.

The two remaining conditions (3.27) and (3.29) follow directly from (3.24) and (3.26), respectively. □

Lemma 3.5 identifies convex sets that contain the spectrum of $2I - \Gamma_r(j\omega), r = 1, 2, 3, 4$, for arbitrary bounded delays. A constructive illustration of the sets $\Omega_r(\omega\mathcal{T})$ for some $\omega\mathcal{T} > 0$ is given in Figure 3.4; numerical computations of these sets are depicted in Figure 3.5. Interestingly, the set containing $\sigma(2I - \Gamma_4(j\omega))$ is smaller than the set containing $\sigma(2I - \Gamma_2(j\omega))$. That is, for MAS with identical delays, the set decreases if the delays are symmetric $\tau_{ji} = \tau_{ij}$ compared to the case of asymmetric delays. For MAS without self-delay or with different self-delay, symmetric delays $\tau_{ij} = \tau_{ji}, T_{ij} = T_{ji}$ result in exactly the same sets as for asymmetric delays, see also Münz et al. (2009c).

The set $\Omega_1(\omega\mathcal{T})$ is a subset of a disc with radius 1 and center +1. The set equals the disc for $\omega\mathcal{T} \geq \pi$. Similarly, the sets $\Omega_2(\omega\mathcal{T}), \Omega_3(\omega\mathcal{T})$, and $\Omega_4(\omega\mathcal{T})$ are subsets of a disc with radius 2 that is centered at +2. The sets equal the disc in all three cases if $\omega\mathcal{T} \geq 2\pi$. In all four cases, the delay bound $\mathcal{T}$ determines how fast the sets increase with ω. This is illustrated exemplarily in Figure 3.6 that shows a numerical computation of the sets $\Omega_1(\omega\mathcal{T})$ for different values of $\omega = \{0.5, 1, 2, 3\}$ and $\mathcal{T} = 0.5$. Moreover, the eigenvalues of a 50×50 ring topology matrix with 150 randomly chosen sets of delays are depicted and the convex hull of the computed eigenvalues is indicated by a thick red line. The delay- and frequency-dependence of the sets Ω_r is a major advantage of the method presented here compared to alternative approaches based on Gershgorin's circle theorem, e.g. Lee and Spong (2006); Tian and Liu (2008); Wang and Elia (2008). From Gershgorin's circle theorem (Horn and Johnson, 1985) we know that the eigenvalues of $D^{-1}A_\tau(j\omega)$ are inside the unit circle. Yet, the set $\Omega_1(\omega\mathcal{T})$ is much more precise about the location of the spectrum of $I - D^{-1}A_\tau(j\omega)$ for small $\omega\mathcal{T}$. Similarly, the sets $\Omega_2(\omega\mathcal{T}), \Omega_3(\omega\mathcal{T})$, and $\Omega_4(\omega\mathcal{T})$ are strictly smaller than the corresponding sets using Gershgorin's circle theorem for small $\omega\mathcal{T}$. Note also that Gershgorin's circle theorem usually provides delay-independent conditions, e.g. Lee and Spong (2006); Wang and Elia (2008). In some cases, consensus cannot be achieved independent of delay, and therefore Gershgorin's circle theorem will not lead to a satisfying result. In contrast, the methodology we present here is able to provide delay-dependent consensus conditions.

At this point, it also becomes clear why our approach applies only to undirected graphs, whereas Gershgorin's circle theorem is also applicable to directed graphs. The normalized adjacency matrix $D^{-1}A$ of an *undirected* graph without delays has real eigenvalues in $[-1, 1]$. Yet, the eigenvalues of $D^{-1}A$ of a *directed* graph without delays are spread all over the unit disc and depend on the graph topology. Hence, similar results as in Lemma 3.5 for directed graphs are very difficult to obtain even without delays. This task becomes even more complicated, if not impossible, for networks with heterogeneous delays. Thus, we restrict this part of our thesis to undirected graphs.

The sets Ω_r contain the spectrum of $2I - \Gamma_r(j\omega)$ for any undirected, arbitrarily large topology and any bounded delays $\tau_{ij}, T_{ij} \in [0, \mathcal{T}]$. Therefore, our subsequent analysis based on these sets provides a robustness with respect to unknown network size and topology as well as unknown but bounded delays. It is of course important to know if these sets and thereby the robustness analysis are very conservative. In Figure 3.5, we show a mathematical computation of the sets $\Omega_r(\omega\mathcal{T})$ for $\omega = 3$ and $\mathcal{T} = 0.5$. Moreover, the eigenvalues of a 50×50 ring topology matrix with 150 randomly chosen sets of delays are depicted and the convex hull of the computed eigenvalues is indicated by a thick red line. We see the sets are not very conservative compared to the location of the eigenvalues. Even more importantly, we can show that the sets Ω_1, Ω_3, and Ω_4 are the smallest convex sets containing all eigenvalues of $2I - \Gamma_r(j\omega), r = 1, 3, 4$ for arbitrary topologies and delays, as explained in the following.

Consider a MAS with two agents with adjacency matrix

$$A = \begin{bmatrix} 0 & 1 \\ 1 & 0 \end{bmatrix}. \tag{3.31}$$

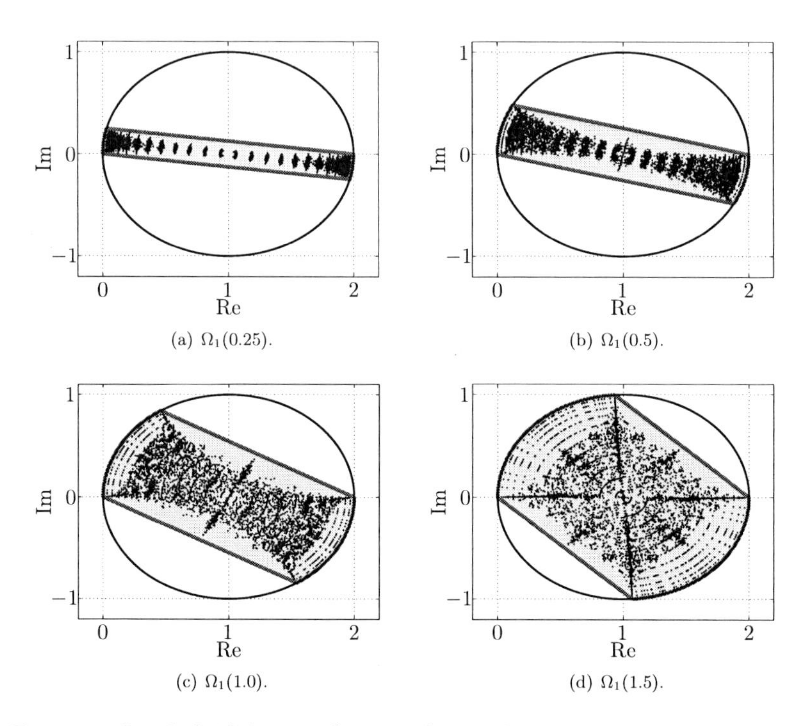

(a) $\Omega_1(0.25)$. (b) $\Omega_1(0.5)$.

(c) $\Omega_1(1.0)$. (d) $\Omega_1(1.5)$.

Figure 3.6: Sets $\Omega_1(\omega\mathcal{T})$ for $\omega = \{0.5, 1, 2, 3\}$ and $\mathcal{T} = 0.5$ and eigenvalues of 50×50 ring topology matrix with 150 randomly chosen sets of delays.

The spectrum of $D^{-1}A = A$ is $\sigma(D^{-1}A) = \{-1, +1\}$. Consider first a MAS without self-delay and choose identical delays $\tau_{ij} = \tau \in [0, \mathcal{T}]$. Under this condition, $2I - \Gamma_1(j\omega) = I + D^{-1}Ae^{-j\omega\tau}$ has eigenvalues $1 \pm e^{-j\omega\tau}$. These eigenvalues are on the boundary of a disc with center 1 and radius 1, i.e. the disc that contains Ω_1. The convex hull of these eigenvalues for all $\tau \in [0, \mathcal{T}]$ is $\Omega_1(\omega\mathcal{T})$.

Similarly, we may proceed for MAS with different self-delay by choosing $\tau_{ij} = \tau \in [0, \mathcal{T}]$ and $T_{ij} = T \in [0, \mathcal{T}]$. For the same topology with adjacency matrix (3.31), we obtain $\sigma(2I - \Gamma_3(j\omega)) = \{2 - e^{-j\omega T} \pm e^{-j\omega\tau}\}$. The convex hull of these eigenvalues for all $\tau, T \in [0, \mathcal{T}]$ is $\Omega_3(\omega\mathcal{T})$.

For MAS with identical self-delay and symmetric delays, we choose again $\tau_{ij} = \tau \in [0, \mathcal{T}]$. For the same topology, we have $\sigma(2I - \Gamma_4(j\omega)) = \{2 - 2e^{-j\omega\tau}, 2\}$. The convex hull of these eigenvalues for all $\tau \in [0, \mathcal{T}]$ results in $\Omega_4(\omega\mathcal{T})$.

The set Ω_2 is the only one that seems conservative, see Figure 3.5(b). There are areas that are contained in Ω_2 but not in the convex hull of the computed eigenvalues. Note however that the there are eigenvalues of $2I - \Gamma_2(j\omega)$ that are not in $\Omega_4(\omega\mathcal{T})$. This justifies the distinction between symmetric and asymmetric delays for MAS with identical self-delay.

3.4.3 Generalized Nyquist Consensus Condition

Now, we have all the preliminary results to formulate a generalized Nyquist consensus condition. Remember that the network return ratio $G_r(j\omega)$ (3.21) is defined such that the open loop system is asymptotically stable. Hence, the characteristic quasi-polynomial (3.20) has all roots in $\mathbb{C}^-$ if the eigenloci of $G_r(j\omega)$ neither touch nor encircle the point -1 as ω goes from $-\infty$ to $+\infty$. The main difficulty is to cope with the roots of $\Delta(s)$ at the origin.

Theorem 3.6 (Generalized Nyquist consensus condition for non-identical agents)**.** *A MAS with agent dynamics $H_i(s)$ (3.1) and delay-interconnection matrix $\Gamma_r(s), r = 1, 2, 3, 4$, (3.5), where $H_i(s)$ satisfy Assumption 3.1, achieves consensus asymptotically on networks of arbitrary size $N \in \mathbb{N}$, with arbitrary connected undirected topologies, and arbitrary heterogeneous bounded delays $\tau_{ij}, T_{ij} \leq \mathcal{T}$ if*

$$-1 \notin -Co\left\{\frac{1}{2 + H_i^{-1}(j\omega)}\Omega_r(\omega\mathcal{T}) : i \in \mathcal{N}\right\}, \tag{3.32}$$

for all $\omega \in \mathbb{R} \setminus \{0\}$, where $\Omega_r(\omega\mathcal{T})$ is defined in Section 3.4.2.

Proof. The generalized Nyquist criterion states that all roots of the characteristic quasi-polynomial are in the open left half plane $\mathbb{C}^-$ if the eigenloci of an asymptotically stable return ratio $G_r(j\omega)$ neither touch nor encircle the point -1 as ω goes from $-\infty$ to $+\infty$. This has been shown in Desoer and Wang (1980) for dynamical systems with rational and non-rational transfer functions. The system considered here belongs to the class of non-rational transfer functions because of the delays. It was shown in Mossaheb (1980) that the assumptions for systems with non-rational transfer functions in Desoer and Wang (1980) are satisfied for time-delay systems like the one considered here.

All roots of $\Delta(s)$ at the origin correspond to exactly one eigenlocus of $G_r(j\omega)$ touching -1 for $\omega = 0$. Recall that

$$G_r(0) = -\text{diag}\left(\frac{\nu_i(0)}{\delta_i(0) + 2\nu_i(0)}\right)(2I - \Gamma_r(0)) = -I + \frac{1}{2}\overline{L}, \quad r = 1, 2, 3, 4,$$

because $\delta_i(0) = 0$. Remember that $\overline{L}$ is positive semidefinite and has a single eigenvalue at the origin if the underlying graph is connected. This single zero eigenvalue of $\overline{L}$ corresponds to a single eigenvalue $\lambda_i(G_r(0)) = -1$, all other eigenvalues of $G_r(0)$ are real and larger than -1 because all nonzero eigenvalues of $\overline{L}$ are strictly positive. Moreover, we have $\det(I + G_r(0)) = \prod_{i=1}^{N}(1 + \lambda_i(G_r(0))) = 0$ where exactly one of the factors $1 + \lambda_i(G_r(0))$ is zero. Thus, one eigenlocus of $G_r(j\omega)$ touches -1 as ω changes from negative to positive values which corresponds to the zero roots of $\Delta_0(s)$. All other eigenloci of $G_r(j\omega)$ neither touch nor encircle the point -1 as ω changes from negative to positive values. It remains to show that the eigenloci of $G_r(j\omega)$ neither touch nor encircle -1 as ω goes from $-\infty$ to 0 and from 0 to $+\infty$. This is true if $-1 \notin \text{Co}\{\sigma(G_r(j\omega)), 0\}$ for all $\omega \neq 0$.

Recall first the following property of the field of values: $\sigma(M_1 M_2) \subset \mathcal{F}(M_1) \times \mathcal{F}(M_2)$ for all $M_1, M_2 \in \mathbb{C}^{N\times N}$ if M_2 is positive semi-definite, see Corollary 1.7.7 in Horn and Johnson (1991). Thereby, $\times$ denotes an elementwise multiplication. In addition, we use Lemma 3.5 to obtain

$$\begin{aligned}
\sigma(G_r(j\omega)) &= -\sigma\left(\text{diag}\left(\frac{\nu_i(j\omega)}{\delta_i(j\omega) + 2\nu_i(j\omega)}\right)(2I - \Gamma_r(j\omega))\right) \\
&= -\sigma\left(\text{diag}\left(\frac{\nu_i(j\omega)}{\delta_i(j\omega) + 2\nu_i(j\omega)}\right)\left(2I - D^{\frac{1}{2}}\Gamma_r(j\omega)D^{-\frac{1}{2}}\right)\right) \\
&\subset -\mathcal{F}\left(\text{diag}\left(\frac{\nu_i(j\omega)}{\delta_i(j\omega) + 2\nu_i(j\omega)}\right)\right) \times \mathcal{F}\left(2I - D^{\frac{1}{2}}\Gamma_r(j\omega)D^{-\frac{1}{2}}\right) \\
&\subset -\text{Co}\left\{\frac{\nu_i(j\omega)}{\delta_i(j\omega) + 2\nu_i(j\omega)} : i \in \mathcal{N}\right\} \times \Omega_r(\omega\mathcal{T}), \\
&= -\text{Co}\left\{\frac{\nu_i(j\omega)}{\delta_i(j\omega) + 2\nu_i(j\omega)}\Omega_r(\omega\mathcal{T}) : i \in \mathcal{N}\right\}, \quad \text{for } r = 1, 2, 3, 4,
\end{aligned}$$

because $2I - D^{\frac{1}{2}}\Gamma_r(s)D^{-\frac{1}{2}}$ is positive semidefinite. Hence, (3.32) guarantees that the eigenloci of the return ratio $G_r(j\omega)$ do not encircle the point -1 as ω goes from $-\infty$ to $+\infty$. Furthermore, the eigenloci do not touch -1 for $\omega \neq 0$. For $\omega = 0$, we have exactly one eigenlocus touching -1 which corresponds to the root of $\Delta(s)$ at the origin as explained above. We conclude that the consensus solutions are asymptotically attracting, i.e. consensus is asymptotically reached. □

Note that condition (3.32) is scalable with the number of delays and with the topology, e.g. the condition is exactly the same for 10 or 1000 delays. However, if there are many different agents H_i, possibly with different number of states n_i, then it might be quite cumbersome to check if (3.32) holds. Therefore, we provide the following equivalent condition illustrated in Figure 3.7:

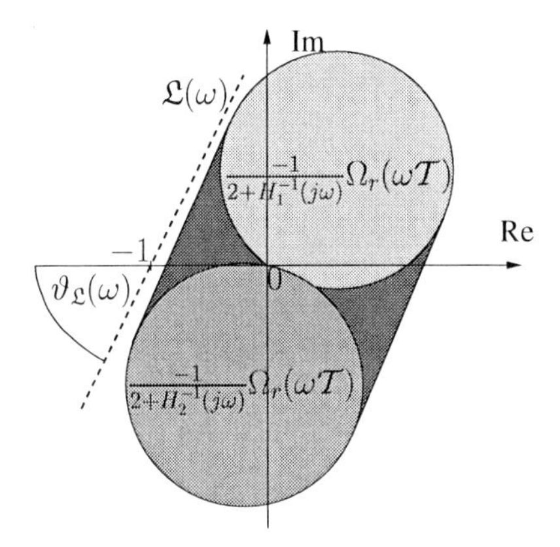

Figure 3.7: Illustation of Equation (3.33) with dashed line $\mathfrak{L}(\omega)$, two circular convex sets $-\frac{1}{2+H_i^{-1}(j\omega)}\Omega_r(\omega\mathcal{T}), i = 1, 2$, and convex hull of these two sets.

Corollary 3.7 (Local generalized Nyquist consensus condition for non-identical agents). *Condition (3.32) is equivalent to the following condition: For every $\omega \in \mathbb{R} \setminus \{0\}$, there exists a $\vartheta_{\mathfrak{L}}(\omega) \in (0, \pi)$ such that*

$$\mathfrak{L}(\omega) \cap -\frac{1}{2 + H_i^{-1}(j\omega)}\Omega_r(\omega\mathcal{T}) = \emptyset, \qquad \text{for all } i \in \mathcal{N}, \tag{3.33}$$

where $\mathfrak{L}(\omega) = \{-1 + \chi e^{j\vartheta_{\mathfrak{L}}(\omega)} : \chi \in \mathbb{R}\}$ is a straight line through -1 with angle $\vartheta_{\mathfrak{L}}(\omega)$ and $\Omega_r(\omega\mathcal{T})$ is defined in Section 3.4.2.

Proof. We show first that (3.32) implies (3.33) for any $\omega \neq 0$. Note that the right hand side of (3.32) is closed for finite any ω. Therefore, we know from Boyd and Vandenberghe (2004), Example 2.20, that there exists a straight line that strictly separates -1 and the convex hull on the right hand side of (3.32) for any $\omega \neq 0$. Denote the angle of this straight line $\vartheta_{\mathfrak{L}}(\omega)$ and note that $\mathfrak{L}(\omega)$ is parallel to this line. Hence, $\mathfrak{L}(\omega) \cap -\mathrm{Co}\left\{\frac{1}{2+H_i^{-1}(j\omega)}\Omega_r(\omega\mathcal{T}) : i \in \mathcal{N}\right\} = \emptyset$ and therefore (3.33) holds, see also Figure 3.7. Since $\Omega_r(\omega\mathcal{T})$ contains the origin for any $\omega\mathcal{T}$, we know that $\vartheta_{\mathfrak{L}}(\omega) \in (0, \pi)$.

Now, we prove the converse, i.e. (3.33) implies (3.32). Since the convex set $\Omega_r(\omega\mathcal{T})$ contains the origin for any $\omega\mathcal{T}$, we know that all $-\frac{1}{2+H_i^{-1}(j\omega)}\Omega_r(\omega\mathcal{T})$ are on the right hand side of $\mathfrak{L}(\omega)$, see Figure 3.7. Note that the convex hull of the right hand side of (3.33) is the right hand side of (3.32). Thus, (3.33) implies that

$$\mathfrak{L}(\omega) \cap -\mathrm{Co}\left\{\frac{1}{2 + H_i^{-1}(j\omega)}\Omega_r(\omega\mathcal{T}) : i \in \mathcal{N}\right\} = \emptyset$$

for all $\omega \neq 0$, i.e. (3.32) holds. □

Corollary 3.7 extends ideas from Lestas and Vinnicombe (2005) where similar consensus conditions for MAS without delays based on a frequency independent line $\mathfrak{L}(\omega) = \mathfrak{L}$ were presented. Note that (3.33) can be separated into local conditions. Given the frequency dependent line $\mathfrak{L}(\omega)$, every agent can check individually if its dynamics $H_i(s)$ satisfy (3.33), without considering the dynamics of the other agents. However, these conditions are not completely local because the angle $\vartheta_{\mathfrak{L}}(\omega)$ of the line $\mathfrak{L}(\omega)$ is a global parameter that has to be determined a priori and that must be known to all agents. Nonetheless, this corollary offers the opportunity for a local controller design, where $\vartheta_{\mathfrak{L}}(\omega)$ is determined a priori and all agents design local controllers in order to satisfy (3.33).

If all agent dynamics are identical, i.e. $H_i(s) = H(s)$, then we can derive the following corollary:

Corollary 3.8 (Generalized Nyquist consensus condition for identical agents). *A MAS with identical agent dynamics $H(s)$ (3.1) and delay-interconnection matrix $\Gamma_r(s), r = 1,2,3,4$, (3.5), where $H(s)$ satisfies Assumption 3.1, achieves consensus asymptotically on networks of arbitrary size $N \in \mathbb{N}$, with arbitrary connected undirected topologies, and arbitrary heterogeneous bounded delays $\tau_{ij}, T_{ij} \leq \mathcal{T}$ if*

$$2 + H^{-1}(j\omega) \notin \Omega_r(\omega\mathcal{T}), \tag{3.34}$$

for all $\omega \in \mathbb{R} \setminus \{0\}$, where $\Omega_r(\omega\mathcal{T})$ is defined in Section 3.4.2.

Proof. Consider (3.32) with $\nu_i(s) = \nu(s)$ and $\delta_i(s) = \delta(s)$. Assume that $\nu(j\omega) \neq 0$ and remember that $\delta(j\omega) + 2\nu(j\omega) \neq 0$ due to Assumption 3.1. We multiply both sides with $2 + H^{-1}(j\omega) = \frac{\delta(j\omega)+2\nu(j\omega)}{\nu(j\omega)}$ and obtain (3.34). For $\nu(j\omega) = 0$, all eigenvalues of $G_{r'}(\omega)$ are zero and $-1 \notin 0$ results immediately. □

Conditions (3.32), (3.33), and (3.34) can be checked graphically or, for particular dynamics, analytically. The latter will be shown in the following sections. Theorem 3.6 and Corollaries 3.7 and 3.8 also provide a simple method to compare the delay robustness of different delay configurations. Therefore, we first compare the sets Ω_r. Note that $(\Omega_1(\omega\mathcal{T}) \cap \mathbb{C}_0^{+j}) \subset \Omega_4(\omega\mathcal{T}) \subset \Omega_2(\omega\mathcal{T}) \subset \Omega_3(\omega\mathcal{T})$ for all $\omega > 0$, where $\mathbb{C}_0^{+j}$ is the closed upper half plane. In other words, the subset of Ω_1 with non-negative imaginary part is contained in all other sets for all $\omega > 0$, see Figure 3.4. Similarly, the subset of Ω_1 with non-positive imaginary part is contained in all other sets for all $\omega < 0$. Moreover, the other sets satisfy $\Omega_4(\omega\mathcal{T}) \subset \Omega_2(\omega\mathcal{T}) \subset \Omega_3(\omega\mathcal{T})$ for all $\omega \neq 0$. This is immediately clear from the fact that MAS with identical self-delay and symmetric delays are a special case of MAS with identical self-delays and possibly asymmetric delays, and this is in turn a special case of MAS with different self-delays.

Consider now agent dynamics such that $\mathrm{Im}\{H_i(j\omega)\} > 0$ for all $i \in \mathcal{N}$ and $\omega > 0$. This implies that $\mathrm{Im}\{(2 + H_i(j\omega))^{-1}\} < 0$ for all $i \in \mathcal{N}$ and $\omega > 0$ and, consequently, the imaginary part of any $z \in \mathrm{Co}\{(2 + H_i(j\omega))^{-1} : i \in \mathcal{N}\}$ is also negative. Therefore, only the upper part $\Omega_r(\omega\mathcal{T}) \cap \mathbb{C}_0^{+j}, r = 1,2,3,4$, is relevant for condition (3.32). We conclude that MAS without self-delay are more robust to delays than MAS with self-delay if $\mathrm{Im}\{H_i(j\omega)\} > 0$ for all $i \in \mathcal{N}$ and $\omega > 0$ because $(\Omega_1(\omega\mathcal{T}) \cap \mathbb{C}_0^{+j}) \subset \Omega_4(\omega\mathcal{T}) \subset$

$\Omega_2(\omega\mathcal{T}) \subset \Omega_3(\omega\mathcal{T})$ for all $\omega > 0$ as derived in the previous paragraph. More robust means in this context that a MAS interconnected without self-delay achieves consensus if the same MAS interconnected with self-delay achieves consensus; however, the inverse is not true in general. The condition $\mathrm{Im}\{H_i(j\omega)\} > 0$ for $\omega > 0$ indeed holds for the most common agent dynamics as for example single integrators and double integrators, see Section 3.5. Further examples where these conditions hold are given in the following sections.

Another important result is that MAS with self-delay cannot reach consensus independent of delay. This becomes clear considering condition (3.34) and taking into account that $H^{-1}(0) = 0$ by Assumption 3.1. Therefore, $2 + H^{-1}(j\omega)$ starts for $\omega = 0$ in the center of the circle that contains $\Omega_r(\omega\mathcal{T}), r = 2, 3, 4$. If the delay bound $\mathcal{T}$ is sufficiently large, these sets $\Omega_r(\omega\mathcal{T})$ cover the complete circle with center 2 and radius 2 for sufficiently small ω where $|H^{-1}(j\omega)| \leq 2$, see Figure 3.4. In these situations, we can always construct topologies and delays such that consensus is not achieved. Thus, MAS with self-delay do not allow for delay-independent consensus conditions. On the contrary, MAS without self-delay may have delay-independent consensus conditions in some cases, see the examples in Sections 3.5 and 3.6.

It is interesting to compare Theorem 3.6 to the results in Lestas and Vinnicombe (2007b,a, 2006), where the authors also derive frequency-dependent sets and set-valued stability and consensus conditions based on these sets. Those results and the results of this thesis build on the generalized Nyquist criterion, yet there are some fundamental differences: In Lestas and Vinnicombe (2007b, 2006), the delay-interconnection matrix $\Gamma_r(s)$ is separated such that $\Gamma_r(s) = B\mathcal{D}(s)B^T$, where B is the incidence matrix and $\mathcal{D}(s)$ is a diagonal matrix containing the delays. Remember that the incidence matrix B of an undirected graph satisfies $L = BB^T$, where L is the Laplacian matrix (Godsil and Royle, 2000). This separation of $\Gamma_r(s)$ requires that the delays are symmetric and that $\Gamma_r(s)$ has zero row sum, i.e. the elements on the diagonal of $\Gamma_r(s)$ equal the row sum of the off diagonal elements multiplied by -1. Considering the different delay configurations studied in this chapter, this condition holds only for MAS with symmetric, identical self-delay, i.e. for Γ_4. Under the assumption that the delay-interconnection matrix can be written as $\Gamma_r(s) = B^T\mathcal{D}(s)B$, set-valued stability and consensus conditions are derived in Lestas and Vinnicombe (2007b,a, 2006). These conditions are very similar to Theorem 3.6, yet the sets Ω_r are replaced by S-hulls of the elements of the diagonal matrix $\mathcal{D}(s)$. This approach gives some additional freedom in the choice of $\mathcal{D}(s)$. Therefore, it is possible to derive local delay-dependent conditions for single integrator MAS. These conditions relate the gain of the non-identical agent dynamics to the delays to their neighbours, i.e. the conditions do not require a global delay bound $\mathcal{T}$. This is the main advantage of the results in Lestas and Vinnicombe (2007b) compared to the results presented here. However, it is not obvious how to derive similar conditions for MAS with asymmetric delays. Moreover, we provide here a unifying framework to compare the delay configuration considered in Lestas and Vinnicombe (2007b) to MAS without self-delay and with different self-delay.

In the following sections of this Chapter, we apply the generalized Nyquist consensus condition to several classes of agent dynamics, e.g. single integrators and relative degree two systems.

3.5 Linear Single Integrator Multi-Agent Systems

First, we investigate the consensus properties of *single integrator MAS*

$$Y_i(s) = \frac{K_i}{s} U_i(s) \tag{3.35}$$

with interconnection (3.3) and coupling gain $K_i \in (0, K], K > 0$. We are interested in the largest K that guarantees consensus for a given delay bound $\mathcal{T}$. The undelayed counterpart of this MAS is the standard model in the literature on consensus and achieves consensus as long as the graph is connected for arbitrary positive K_i, e.g. Ren and Beard (2008). Linear single integrator models with delayed feedback have been studied for example in Bliman and Ferrari-Trecate (2008); Olfati-Saber and Murray (2004); Papachristodoulou and Jadbabaie (2005). The following corollaries state consensus conditions for single integrator MAS. The proofs are given in Appendix C.1 and C.2.

Corollary 3.9 (Single integrator MAS without self-delay). *A single integrator MAS* (3.35) *without self-delay* (3.3a) *with gain $K_i > 0$, arbitrary size $N \in \mathbb{N}$, arbitrary delays $\tau_{ij} \leq \mathcal{T}$, and arbitrary connected topology achieves consensus asymptotically independent of delay, i.e. for any $\mathcal{T}$ and $K_i > 0$.*

Corollary 3.10 (Single integrator MAS with self-delay). *A single integrator MAS* (3.35) *with identical self-delay* (3.3b) *or different self-delay* (3.3c), *with gain $K_i \in (0, K]$, arbitrary size $N \in \mathbb{N}$, arbitrary delays $\tau_{ij}, T_{ij} \leq \mathcal{T}$, and arbitrary connected topology achieves consensus asymptotically if*

$$K < \frac{\pi}{4\mathcal{T}}. \tag{3.36}$$

It is worth mentioning that this robustness condition is exact for arbitrary delays and topologies. In other words, if $K \geq \frac{\pi}{4\mathcal{T}}$, then we can construct topologies and delays such that consensus in not achieved. This is shown briefly in the following: Choose identical delays $T_{ij} = \tau_{ij} = \tau \leq \mathcal{T}$, sufficiently large such that $\frac{\pi}{4\tau} \leq K$, and a topology consisting of two interconnected agents with adjacency matrix as in (3.31). Thus, we have

$$\Gamma_2(s) = \Gamma_3(s) = \Gamma_4(s) = D^{-1} L e^{-s\tau} = \begin{bmatrix} 1 & -1 \\ -1 & 1 \end{bmatrix} e^{-s\tau}.$$

The spectrum of $\Gamma_2(s)$ is $\{0, 2e^{-s\tau}\}$. In this case, the characteristic quasi-polynomial of the closed loop system is

$$\Delta(s) = \det\left(sI + K\Gamma_2(s)\right) = s(s + 2Ke^{-s\tau}).$$

Clearly, $\Delta(s)$ has a root at the origin which corresponds to the consensus solutions. We further analyse the term in parenthesis, which reduces to a Hurwitz polynomial as

$\tau \to 0$, i.e. $s+2K$. This implies that all nonzero roots of $\Delta(s)$ are in $\mathbb{C}^-$ for small delays τ. We detect the first nonzero roots of $\Delta(s)$ on the imaginary axis by setting $s = j\omega$

$$s + 2Ke^{-s\tau}\Big|_{s=j\omega} = j\omega + 2Ke^{-j\omega\tau} = 0,$$

see Appendix A and Gu et al. (2003) for details. This equality can be solved by distinguishing modulo and phase. We obtain $\omega = 2K$ and $\omega\tau = \frac{\pi}{2}(+l2\pi), l = 0, 1, 2, \ldots$. Eliminating ω leads to $K = \frac{\pi}{4\tau}$. For any K, we may choose $\tau = \frac{\pi}{4K}$ satisfying $\tau \leq \mathcal{T}$ because of our assumption $K \geq \frac{\pi}{4\mathcal{T}}$ above. In summary, the characteristic quasi-polynomial of this MAS with $\tau = \frac{\pi}{4K}$ has nonzero roots on the imaginary axis, i.e. consensus in not reached for all initial conditions in this case.

Comparing Corollary 3.9 and 3.10 to previous publications, we know that single integrator MAS without self-delay, i.e. (3.35),(3.3a), achieve consensus independent of delay, see Papachristodoulou and Jadbabaie (2006); Lee and Spong (2006) and Section 4.3.3. This has been shown using Lyapunov-Razumikhin functions and Gershgorin's circle theorem, respectively. Here, this result is recovered in Corollary 3.9 using the generalized Nyquist consensus condition. For MAS with identical self-delay (3.3b), it has been shown in Olfati-Saber and Murray (2004) using also the Nyquist criterion that consensus is reached under exactly the same condition if all delays are identical, i.e. $\tau_{ij} = T_{ij} = \tau$. In Bliman and Ferrari-Trecate (2008); Lestas and Vinnicombe (2007b), the same result is obtained for heterogeneous symmetric delays $\tau_{ij} = \tau_{ji}, T_{ij} = T_{ji}$. Corollary 3.10 extends this result to MAS with heterogeneous, asymmetric delays and even to MAS with different self-delays. Interestingly, the delay bound is the same in all cases with self-delay. Corollary 3.10 also nicely illustrates the relation between the upper bound on the coupling gain K and the delay bound $\mathcal{T}$. As the maximal delay $\mathcal{T}$ increases, the maximal coupling gain K has to decrease in order to maintain the consensus properties of the MAS.

3.6 Relative Degree Two Multi-Agent Systems

In this section, we consider MAS with relative degree two. More precisely, we investigate identical agent dynamics

$$Y_i(s) = \frac{K}{s(s+\rho)} U_i(s), \tag{3.37}$$

with gain $K > 0$ and damping $\rho > 0$. We study these agent dynamics in order to compare the results to conditions in Chapter 5, where we will investigate nonlinear agents with relative degree two.

We start with a condition for MAS without self-delay which is proven in Appendix C.3:

Corollary 3.11 (Relative degree two MAS without self-delay)**.** *A relative degree two MAS* (3.37) *without self-delay* (3.3a) *with gain $K > 0$ and damping $\rho > 0$, arbitrary size $N \in \mathbb{N}$, arbitrary delays $\tau_{ij} \leq \mathcal{T}$, and arbitrary connected topology achieves consensus asymptotically independent of delay if*

$$\rho^2 \geq 2K. \tag{3.38}$$

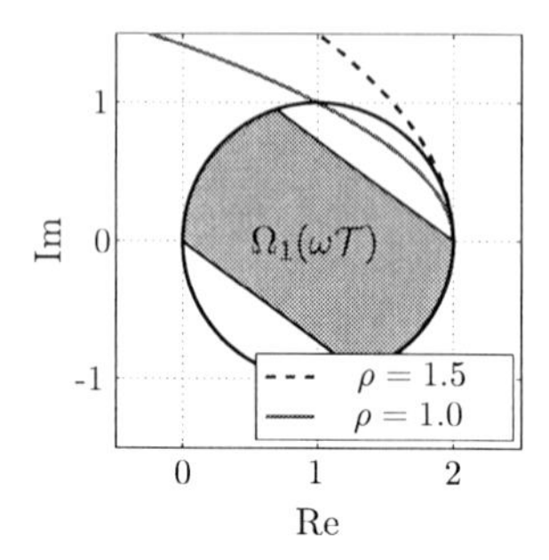

Figure 3.8: Illustration of $\Omega_1(\omega\mathcal{T})$, (3.23a), for $\omega = 2.5$ and $\mathcal{T} = 0.5$ and $2 + H^{-1}(j\omega)$ with $H(s)$ given in (3.37) with $K = 1$ for $\rho = 1$ (red solid line) and $\rho = 1.5$ (blue dashed line), respectively.

For $\rho^2 < 2K$, this MAS achieves consensus asymptotically if

$$\mathcal{T} < \frac{2}{\sqrt{2K - \rho^2}} \arctan \frac{\rho}{\sqrt{2K - \rho^2}} \tag{3.39}$$

Two exemplary transfer functions H as in (3.37) with $K = 1, \rho \in \{1, 1.5\}$ are illustrated in Figure 3.8. The MAS with $\rho = 1.5$ satisfies $\rho^2 > 2K$, i.e. the dashed (blue) curve does not enter the disc that contains Ω_1. Hence, this MAS reaches consensus independent of delay. For the MAS with $\rho = 1$, consensus is only guaranteed for sufficiently small delays because $\rho^2 \leq 2K$, i.e. the solid (red) line enters the disk. In this example, consensus is guaranteed if

$$\mathcal{T} < 2\arctan(1) = \frac{\pi}{2}.$$

It is interesting to see that the robustness condition (3.39) is exact. That is, if $\mathcal{T}$ is such that the right and left hand side of (3.39) are equal, then there exist topologies and delay values such that consensus is not reached. One possible choice are identical delays $\tau_{ij} = \mathcal{T}$ and a MAS with four agents in star topology with adjacency matrix

$$A = \begin{bmatrix} 0 & 1 & 1 & 1 \\ 1 & 0 & 0 & 0 \\ 1 & 0 & 0 & 0 \\ 1 & 0 & 0 & 0 \end{bmatrix}. \tag{3.40}$$

The spectrum of $D^{-1}A$ is $\sigma(D^{-1}A) = \{-1, 0, 0, +1\}$. The eigenvalue -1 and the identical delays $\tau_{ij} = \mathcal{T}$ guarantee that $2I - \Gamma_1(j\omega)$ has in fact an eigenvalue at $1 - e^{-\omega\mathcal{T}}$. Moreover, straightforward calculations show that $2 + H^{-1}(j\omega) = 1 - e^{-\omega\mathcal{T}}$ for $\omega = \sqrt{2K - \rho^2}$ and $\mathcal{T} = \frac{2}{\sqrt{2K-\rho^2}} \arctan \frac{\rho}{\sqrt{2K-\rho^2}}$. Therefore, $-\frac{1}{2+H^{-1}(j\omega)}(2I - \Gamma_1(j\omega))$ has an eigenvalue -1

under these conditions, see also the proof of Corollary 3.11 in Appendix C.3. We conclude that the characteristic quasi-polynomial $\Delta(s)$ has a nonzero root at $j\omega = j\sqrt{2K - \rho^2}$ on the imaginary axis. Thus, consensus is not reached in this case.

Now, we turn to MAS with self-delay, where we have the following result, which is proven in Appendix C.4:

Corollary 3.12 (Relative degree two MAS with self-delay). *A relative degree two MAS* (3.37) *with identical self-delay* (3.3b) *or different self-delay* (3.3c) *with gain $K > 0$ and damping $\rho > 0$, arbitrary size $N \in \mathbb{N}$, arbitrary delays $\tau_{ij}, T_{ij} \leq \mathcal{T}$, and arbitrary connected topology achieves consensus asymptotically if*

$$\mathcal{T} < \frac{\rho}{2K} \tag{3.41}$$

In contrast to Corollaries 3.9 to 3.11, Corollary 3.12 does not solve (3.34) exactly but introduces some additional conservatism, see the proof in Appendix C.4. There are in most cases larger delay bounds $\mathcal{T}$ violating (3.41) for which (3.34) still holds. We present this conservative result in order to show how to apply Theorem 3.6 and Corollary 3.8 to more complicated, high order agent dynamics $H(s)$. For such MAS, it is usually very cumbersome or even impossible to transform the set-valued conditions (3.32), (3.33), and (3.34) into an exact delay bound like (3.39). In these cases, it might be appropriate to introduce conservatism in order to achieve simpler conditions. This idea will be further exploited in Section 3.7.

As in Section 3.5, we see in both Corollary 3.11 and 3.12 that larger delays require smaller coupling gains. Yet, for relative degree two agent dynamics as in (3.37), the damping ρ also plays an important role in the delay dependent conditions (3.39) and (3.41): Larger damping allows for larger delays, see also Chapter 5.

3.7 Consensus Controller Design

In this section, we present a delay-dependent consensus controller design for a class of linear MAS (3.1). We consider agents that consist of a series interconnection of a plant $P(s)$ and a consensus controller $C(s)$, such that the networks output U_i is input of the controller C. Thus, the open loop transfer function from U_i to Y_i is

$$Y_i(s) = P(s)C(s)U_i(s) = H(s)U_i(s). \tag{3.42}$$

We have the following assumption on the plant P:

Assumption 3.13. *The plant transfer function $P(s)$ is proper, minimum phase, and has all poles in the open left half plane except for a possible single pole at the origin.*

Assumption 3.13 is not very restrictive taking into account that $H(s)$ has to satisfy Assumption 3.1. Remember that all coefficients of the denominator polynomial of a stable transfer function P are either non-negative or non-positive. The same holds for all coefficients of the numerator polynomial of a minimum phase transfer function P.

For simplicity of presentation, we assume that all coefficients of the numerator and denominator polynomial of P are non-negative. If this is not the case, e.g. if the coefficients of the numerator are non-positive and the coefficients of the denominator are non-negative, our subsequent controller design is still valid if $-C(s)$ instead of $C(s)$ is used.

We want to design a controller C that achieves consensus for arbitrary delays with known upper bound $\mathcal{T}$. The controller shall satisfy the following properties:

Design Condition 3.14. *The controller C is proper, stable, and minimum phase with non-negative coefficients in the numerator and denominator polynomial. Moreover, C is such that $H = PC$ has exactly one pole at the origin and is strictly proper.*

We first present the fundamental idea of the design. Consider $H^{-1}(j\omega) = (P(j\omega)C(j\omega))^{-}$ for $\omega \ll \frac{\pi}{\mathcal{T}}$. For these frequencies, the sets $\Omega_r, r = 1, 2, 3, 4$, are strict subsets of the discs that enclose Ω_r. With $H(s) = \frac{\nu_0+\nu_1 s+\ldots+\nu_{\tilde{n}} s^{\tilde{n}}}{\delta_0+\delta_1 s+\ldots+\delta_n s^n}$ and $\delta_0 = 0$, we have

$$\begin{aligned} H^{-1}(j\omega) &= \frac{j\omega\delta_1 - \omega^2\delta_2 - j\omega^3\delta_3 + \ldots}{\nu_0 + j\omega\nu_1 - \omega^2\nu_2 - j\omega^3\nu_3 + \ldots} \\ &\approx \frac{\omega^2(\delta_1\nu_1 - \delta_2\nu_0) + j\omega(\delta_1\nu_0 + \omega^2(\nu_1\delta_2 - \nu_0\delta_3 - \nu_2\delta_1)) + \mathcal{O}_4(j\omega)}{|\nu_0 + j\omega\nu_1 + \ldots + (j\omega)^{\tilde{n}}\nu_{\tilde{n}}|^2}, \end{aligned} \tag{3.43}$$

for sufficiently small $\omega \ll \frac{\pi}{\mathcal{T}}$ where $\mathcal{O}_4(j\omega)$ denotes higher order terms in $j\omega$ at least of order 4. We know that δ_1, ν_0 are strictly positive and $\delta_2, \delta_3, \nu_1, \nu_2$ are non-negative because of Assumption 3.13 and Design Condition 3.14. From (3.43), we see that $\mathrm{Im}\{H^{-1}(j\omega)\} > 0$ and $\mathrm{Im}\{H^{-1}(j\omega)\} > \mathrm{Re}\{H^{-1}(j\omega)\}$ for sufficiently small $\omega > 0$. Hence, $2 + H^{-1}(j\omega)$ starts at 2 for $\omega = 0$ and varies toward $2 + j$ for sufficiently small $\omega > 0$, i.e. the Nyquist plot is orthogonal to the real axis at $+2$ as for example in Figure 3.8. This property is crucial for our controller design.

Now, we distinguish the two fundamentally different cases without self-delay, i.e. interconnection (3.3a), and with self-delay, i.e. interconnection (3.3b) and (3.3c). For the first case, it is quite easy to design a controller such that consensus is reached. Note that the elements of Ω_1 shifted by 1 to the left are contained in the unit circle. With some abuse of notation, we refer to this set as $\Omega_1 - 1$. Hence, C has to be designed such that $|1 + (P(j\omega)C(j\omega))^{-1}| > 1$ for all $\omega \neq 0$, see the dashed (blue) Nyquist plot in Figure 3.8 for an example. Then, we may conclude directly that $2 + (P(j\omega)C(j\omega))^{-1} \notin \Omega_1(\omega\mathcal{T})$ for any $\omega \neq 0$, i.e. the designed controller achieves consensus independent of delay, see Corollary 3.8. The same design condition has been proposed in Lee and Spong (2006) based on Gershgorin's circle theorem.

The more challenging case are MAS with self-delay because the controller design is delay-dependent. This becomes clear if we consider Figure 3.4 and recall that $2 + H^{-1}(0) = 2$. Thus, Ω_2, Ω_3, and Ω_4 cover the complete disc around $+2$ with radius 2 for sufficiently large delay bounds $\mathcal{T}$ before the transfer function $2 + H^{-1}(j\omega)$ has left this disc. As a result, the controller design for MAS with self-delay cannot be delay-independent, remember also the discussion in Section 3.4.3. For this case, Algorithm 3.15

designs a delay-dependent controller C for a given delay bound $\mathcal{T}$, such that $2+H^{-1}(j\omega)$ leaves the disc enclosing the sets $\Omega_r, r=2,3,4$, sufficiently fast.

Algorithm 3.15 (Consensus control of stable, minimum phase agents with self-delay). *This algorithm constructs a consensus controller $C(s)$ satisfying Design Condition 3.14 for a linear MAS consisting of agents with dynamics $P(s)$ satisfying Assumption 3.13 with interconnection* (3.3b) *or* (3.3c) *with bounded delays $\tau_{ij}, T_{ij} \leq \mathcal{T}$.*

1. *Construct a polynomial $c(s) = \prod_{i=1}^{n_c}(s+c_i)$ for some $n_c \in \mathbb{N}$ and $c_i \in \mathbb{R}$ such that*
   - *$c_i \geq 0, i=1,\ldots,n_c$, and $P(s)c^{-1}(s)$ has a single pole at the origin,*
   - *$|P^{-1}(j\omega)c(j\omega)|$ is non-decreasing with ω.*

   For instance, choose c_i slightly smaller than the break point frequencies[1] *of the zeros of P.*

2. *Find a sufficiently small $c_0 > 0$ such that*

$$\mathrm{Im}\{P^{-1}(j\omega)c_0^{-1}c(j\omega)\} > 2\sin(\omega\mathcal{T}) \tag{3.44}$$

   for all $\omega > 0$ where $|P^{-1}(j\omega)c_0^{-1}c(j\omega)| < 2$.

The consensus controller is $C(s) = c_0 c^{-1}(s)$.

The first step ensures that $P(s)C(s)$ has a single root at the origin and $|P(j\omega)C(j\omega)|$ is non-increasing. The second step guarantees that $2+H^{-1}(j\omega)$ leaves the disc enclosing Ω_r sufficiently fast such that $2+H^{-1}(j\omega) \notin \Omega_r(\omega\mathcal{T}), r=2,3,4$. Note that $\mathrm{Im}\{z\} \leq 2\sin(\omega\mathcal{T})$ for all $z \in \Omega_r(\omega\mathcal{T})$ and $\omega < \frac{\pi}{2\mathcal{T}}$, see Figure 3.4 or Figure C.1(a) in Appendix C for an illustration. Since $\mathrm{Im}\{H^{-1}(j\omega)\} > 0$ and $\mathrm{Im}\{H^{-1}(j\omega)\} > \mathrm{Re}\{H^{-1}(j\omega)\}$ for sufficiently small $\omega > 0$, see (3.43), it is always possible to find a c_0 for any $\mathcal{T}$. Note that c_0 can be considered as the coupling gain of the consensus controller, and again we have to take a smaller coupling gain c_0 for a larger delay bound $\mathcal{T}$ in (3.44).

Algorithm 3.15 is simple yet conservative. The conservatism can be easily reduced using the following ideas. Some of the controller poles c_i are actually not necessary because $|H^{-1}(j\omega)|$ may in fact decrease for larger frequencies as long as $|H^{-1}(j\omega)| > 2$. In the second step, the bound $2\sin(\omega\mathcal{T})$ is also conservative, in particular if $H^{-1}(j\omega)$ is in the first quadrant for small ω, i.e. if $\delta_1\nu_1 > \delta_2\nu_0$, see (3.43). In this case, we may replace $2\sin(\omega\mathcal{T})$ by $\sin(\omega\mathcal{T})\left(1+\frac{1-\cos(\omega\mathcal{T})}{1+\cos(\omega\mathcal{T})}\right)$, for example, see also Figure C.1(a) on page 116.

We illustrate this controller design in an example.

Example 3.16 (Consensus controller design for MAS with self-delay). *Consider a MAS with agent dynamics $P(s) = \frac{s+2}{s^2+s}$. The number of agents as well as the exact topology are unknown. We assume that the topology is undirected and connected. We want to*

[1] Break point frequencies are used in the straight line approximations of Bode plots, see for example Skogestad and Postlethwaite (2004)

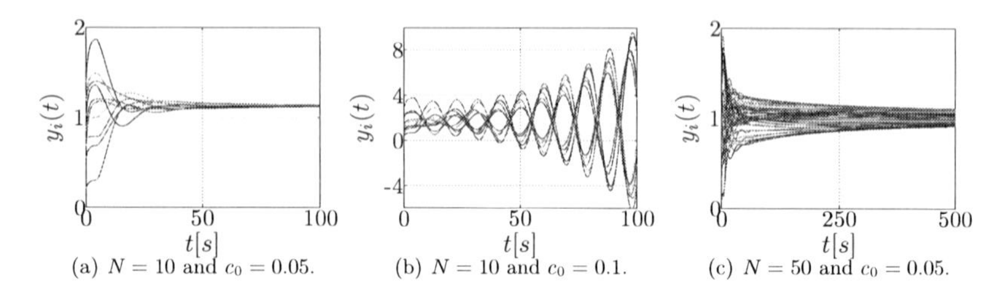

(a) $N = 10$ and $c_0 = 0.05$. (b) $N = 10$ and $c_0 = 0.1$. (c) $N = 50$ and $c_0 = 0.05$.

Figure 3.9: Simulation results for MAS with identical self-delay with dynamics $P(s) = \frac{s+2}{s^2+s}$, ring topology, and random delays $\tau_{ij} \leq 5$ for different numbers of agents N and different values of c_0, see Example 3.16.

design a controller C that guarantees consensus for interconnections with different self-delay (3.3c) with bounded delays $\tau_{ij}, T_{ij} \leq \mathcal{T}$.

The first step of Algorithm 3.15 requires to construct polynomials $c(s)$ such that $P(j\omega)c^{-1}(j\omega)$ has a pole at the origin and $|P^{-1}(j\omega)c(j\omega)|$ is non-decreasing with ω. Clearly, P has a pole at the origin, i.e. $c(s)$ may not have a root at the origin. In order to investigate the magnitude of $P^{-1}(j\omega)$, we compute

$$P^{-1}(j\omega) = \frac{j\omega - \omega^2}{2 + j\omega} = \frac{-\omega^2 + j\omega(2 + \omega^2)}{4 + \omega^2}.$$

We see that the magnitude of $P^{-1}(j\omega)$ is strictly increasing with ω. Therefore, we may choose $c(s) = 1$, i.e. a static controller $C(s) = c_0$ achieves consensus for a suitably chosen c_0.

In the second step, we determine an upper bound $\bar{c}_0$ such that consensus is achieved for all $c_0 \leq \bar{c}_0$. Therefore, we have to compare the imaginary part of $\bar{c}_0^{-1}P^{-1}(j\omega)$ to $2\sin(\omega\mathcal{T})$, i.e.

$$\frac{1}{\bar{c}_0} > 2\mathcal{T}\frac{\sin\omega\mathcal{T}}{\omega\mathcal{T}}\frac{4 + \omega^2}{2 + \omega^2}$$

guarantees consensus. Note that $\frac{\sin\omega\mathcal{T}}{\omega\mathcal{T}} < 1$ and $\frac{4+\omega^2}{2+\omega^2} < 2$ for all $\omega \neq 0$. Therefore, $c_0 \leq \bar{c}_0 = \frac{1}{4\mathcal{T}}$ guarantees consensus.

Simulation results are shown in Figure 3.9 for a MAS with identical self-delay with dynamics $P(s) = \frac{s+2}{s^2+s}$. The topology is an undirected ring. The asymmetric delays are randomly chosen such that $\tau_{ij} \leq \mathcal{T} = 5$. The constant initial conditions are also randomly chosen. Figure 3.9(a) shows consensus for a MAS with $N = 10$ agents and $c_0 = \bar{c}_0 = 0.05$. These results are compared to the simulation of the same MAS with controller gain $c_0 = 0.1$, see Figure 3.9(b). In this simulation consensus is clearly not reached. We ran further simulations with $c_0 = 0.1$ with different random delay values. In some simulations, consensus is actually reached, which is not shown here. This is not surprising because Corollary 3.8 is quite accurate for arbitrary topologies and delays, *but for specific topologies and delays, consensus can be reached even if* (3.34) *is violated. Figure 3.9(c) shows the simulation results of a MAS with* $N = 50$ *agents and* $c_0 = 0.05$. *This simulation nicely illustrates the scalability of the controller design for large MAS. We see however that consensus is reached much slower than for the MAS with* 10 *agents. This is due to weaker connectivity of the underlying ring topology with increasing number of agents. The connections between the connectivity of the underlying graph and the convergence rate are subject of the next section. Summarizing, this simulation example shows the effectiveness of the proposed controller design algorithm.*

3.8 Convergence Rate of Linear Single Integrator Multi-Agent Systems

Now, we use the generalized Nyquist criterion to derive an lower bound on the convergence rate of identical single integrator MAS with heterogeneous delays. The convergence rate describes how fast the solutions approach consensus. Convergence has been studied thoroughly for MAS without delays, e.g. Olfati-Saber (2005). The convergence rate of discrete-time MAS with heterogeneous delays has been investigated in Nedić and Ozdaglar (2010). In continuous-time, Sun and Lemmon (2007) study the convergence rate of MAS with homogeneous delays using a Padé approximation of the delay. A numerical computation of the rightmost roots of the characteristic quasi-polynomial is proposed in Scutari et al. (2008). This is however not a scalable convergence rate analysis because the number of agents and the topology must be known in order to compute the convergence rate. Here, we present a scalable convergence rate conditions for continuous-time single integrator MAS with heterogeneous feedback delays on arbitrary topologies where only a lower bound on the algebraic connectivity of the graph is known.

3.8.1 Robust, Scalable Convergence Rate Conditions

We first present the fundamental idea. Consider a single integrator MAS (3.35) with identical agent dynamics $H = H_i$ and undelayed feedback. The corresponding characteristic equation is

$$\Delta_0(s) = \det\left(sI + K(I - D^{-1}A)\right) = 0, \quad K > 0,$$

see also (3.7). If the underlying network of the MAS is connected, Δ_0 has a single root at the origin and all nonzero roots are in the open left half plane. The root at the origin corresponds to the consensus solutions. The convergence rate toward the consensus solutions is determined by the rightmost nonzero root of Δ_0. Since $\overline{L} = I - D^{-1}A$ is symmetric, the rightmost nonzero root of Δ_0 is $-\lambda_2 K$, where $\lambda_2 > 0$ is the second smallest eigenvalue of $\overline{L}$. Recall that λ_2 is the algebraic connectivity of the undirected graph, i.e. it is a measure of how well the graph is connected, see Appendix B. For any $\psi \in (0, \lambda_2 K)$ we define $\Psi = \{s \in \mathbb{C} : \mathrm{Re}\{s\} < -\psi\}$. Hence, all nonzero roots of Δ_0 are in Ψ and a lower bound on the convergence rate of the undelayed MAS is ψ. This is illustrated in Figure 3.10(a), where the red circles ($\circ$) indicate exemplary roots of Δ_0.

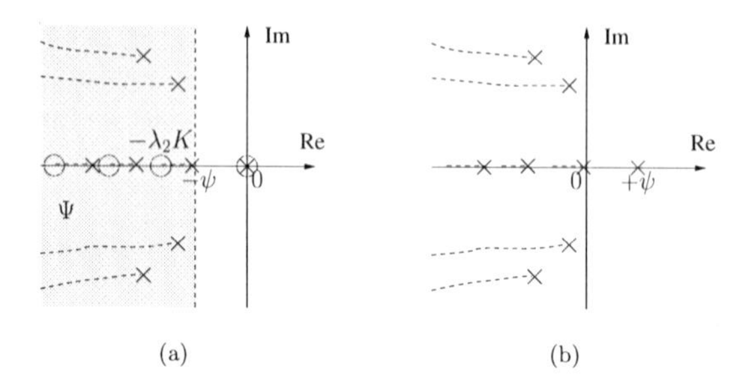

Figure 3.10: Roots of characteristic polynomial Δ_0 (denoted ○) and characteristic quasi-polynomial Δ (denoted ×) for particular $\tau_{ij} < \mathcal{T}$ (left) and corresponding roots of $\Delta(s-\psi)$ (right).

Consider now the characteristic quasi-polynomial of the single integrator MAS (3.35) with interconnection (3.3) and with $K > 0$

$$\Delta(s) = \det\left(sI + K\Gamma_r(s)\right). \tag{3.45}$$

As discussed in Section 3.4.1, there is a single root of Δ at the origin if the graph is connected. As for undelayed MAS, this root corresponds to the consensus solutions. All nonzero roots of Δ are in the open left half plane if the conditions of Corollary 3.9 and 3.10 are satisfied. In this subsection, we will provide additional conditions for $\mathcal{T}$ and K such that all nonzero roots of Δ are in Ψ for all $\tau_{ij}, T_{ij} \leq \mathcal{T}$. This implies that consensus is reached exponentially with convergence rate ψ, see Appendix A. In the case of MAS without self-delay, we have shown in Corollary 3.9 that consensus is achieved independent of delay. Yet, consensus can be achieved arbitrarily slow for large delays. The delay-dependent condition in Theorem 3.17 below is much more useful if we want to guarantee a fast convergence, see also Subsection 3.8.2 below.

A key property in our analysis is the continuity of the roots of Δ with respect to the delays τ_{ij}, T_{ij}, see Appendix A. In fact, as $\mathcal{T} \to 0$, a finite number of roots of Δ converge toward the roots of Δ_0 and the real part of all remaining roots goes to $-\infty$, see Michiels and Niculescu (2007). Conversely, for sufficiently small $\mathcal{T}$, the roots of Δ remain in Ψ because they are either close to the roots of Δ_0 or "close" to $-\infty$. This is illustrated in Figure 3.10(a), where the blue crosses (×) indicate roots of Δ for particular $\tau_{ij}, T_{ij} < \mathcal{T}$. The dashed lines show the paths on which the roots have moved as τ_{ij}, T_{ij} have increased. These root paths are not unique because there are several ways to increase τ_{ij}, T_{ij}. The major challenge is to derive conditions on the delay bound $\mathcal{T}$ such that the real part of the nonzero roots of Δ remains smaller than $-\psi$. Therefore, we shift all roots of Δ by ψ to the right, i.e. we define a new polynomial $\tilde{\Delta}(s) = \Delta(s-\psi)$, see Figure 3.10(b). The zero root of Δ corresponds to a root of $\tilde{\Delta}$ at $+\psi$ in the right half plane $\mathbb{C}^+$. If all other roots of $\tilde{\Delta}$ are in the left half plane, then we know that all nonzero roots of Δ are in Ψ and the MAS has convergence rate ψ.

We test the location of the roots using the eigenloci of the return ratios $G_r(s)$ (3.21) for $s = -\psi + j\omega$. If these eigenloci encircle the point -1 exactly once clockwise, i.e. the Nyquist criterion certifies a single root of $\tilde{\Delta}$ in $\mathbb{C}^+$ (Skogestad and Postlethwaite, 2004; Mossaheb, 1980), then we know that all nonzero roots of Δ are in Ψ. This root of $\tilde{\Delta}$ in $\mathbb{C}^+$ is the zero root of Δ. These ideas lead to the following theorems, which are proven in Appendix C.5, C.6, and C.7:

Theorem 3.17 (Convergence rate of single integrator MAS without self-delay). *A single integrator MAS (3.35) of arbitrary size $N \in \mathbb{N}$, with identical dynamics $H_i(s) = \frac{K}{s}$, with interconnection without self-delay (3.3a), gain $K \in (0, \frac{1}{\mathcal{T}})$, and arbitrary symmetric delays $\tau_{ij} = \tau_{ji} \leq \mathcal{T}$ achieves consensus exponentially with convergence rate $\psi \in (0, \overline{\lambda}_2 K)$, satisfying*

$$e^{\psi \mathcal{T}} < \min\left\{1 - \frac{\psi}{K} + \overline{\lambda}_2, \frac{1}{K\mathcal{T}}\right\}, \tag{3.46}$$

where $\overline{\lambda}_2 \in (0,1)$ is a lower bound on the second smallest eigenvalue λ_2 of $\overline{L} = I - D^{-1}A$.

Theorem 3.18 (Convergence rate of single integrator MAS with identical self-delay). *A single integrator MAS (3.35) of arbitrary size $N \in \mathbb{N}$, with identical dynamics $H_i(s) = \frac{K}{s}$, with interconnection with identical self-delay (3.3b), gain $K \in (0, \frac{1}{2\mathcal{T}})$, and arbitrary symmetric delays $\tau_{ij} = \tau_{ji} \leq \mathcal{T}$ achieves consensus exponentially with convergence rate $\psi \in (0, \overline{\lambda}_2 K)$, satisfying*

$$e^{\psi \mathcal{T}} < \frac{1}{2K\mathcal{T}}, \tag{3.47}$$

where $\overline{\lambda}_2 \in (0,1)$ is a lower bound on the second smallest eigenvalue λ_2 of $\overline{L} = I - D^{-1}A$.

Theorem 3.19 (Convergence rate of single integrator MAS with different self-delay). *A single integrator MAS (3.35) of arbitrary size $N \in \mathbb{N}$, with identical dynamics $H_i(s) = \frac{K}{s}$, with interconnection with different self-delay (3.3c), gain $K \in (0, \frac{1}{2\mathcal{T}})$, and arbitrary symmetric delays $\tau_{ij} = \tau_{ji} \leq \mathcal{T}, T_{ij} = T_{ji} \leq \mathcal{T}$ achieves consensus exponentially with convergence rate $\psi \in (0, \overline{\lambda}_2 K)$, satisfying*

$$e^{\psi \mathcal{T}} < \min\left\{1 - \frac{\psi}{K} + \overline{\lambda}_2, \frac{1}{2K\mathcal{T}}\right\}, \tag{3.48}$$

where $\overline{\lambda}_2 \in (0,1)$ is a lower bound on the second smallest eigenvalue λ_2 of $\overline{L} = I - D^{-1}A$.

As with most convergence rate conditions for MAS without delays, e.g. Olfati-Saber (2005), conditions (3.46), (3.47), and (3.48) are not independent of the topology but depend on the algebraic connectivity of the underlying graph represented by λ_2. These conditions are robust to unknown topologies as long as their algebraic connectivity is larger than a lower bound, i.e. $\lambda_2 \geq \overline{\lambda}_2$. For any topology satisfying this inequality, Theorems 3.17, 3.18, and 3.19 provide a lower bound on the convergence rate. On the other hand, the guaranteed convergence rate ψ is non-increasing for increasing delays $\mathcal{T}$ because we require a robustness for arbitrary delays $\tau_{ij}, T_{ij} \leq \mathcal{T}$.

There always exists a sufficiently small $\psi > 0$ that satisfies (3.46) if $\mathcal{T}K < 1$ because $1 - \frac{\psi}{K} + \overline{\lambda}_2 > 1$ for sufficiently small ψ. Similarly, there always exists a sufficiently small $\psi > 0$ that satisfies (3.47), (3.48) if $2\mathcal{T}K < 1$. Comparing the results of Theorem 3.17 and 3.18, we see that (3.47) is more restrictive than (3.46) if $1 - \frac{\psi}{K} + \overline{\lambda}_2 > \frac{1}{2K\mathcal{T}}$. Conversely, if $1 - \frac{\psi}{K} + \overline{\lambda}_2 < \frac{1}{2K\mathcal{T}}$, then (3.46) is more restrictive than (3.47). Note however that ψ also appears on the left hand side of (3.46), (3.47). Thus, a clear statement without restricting $\mathcal{T}K$ is not possible. Clearly, both (3.46) and (3.47) hold if condition (3.48) is satisfied, i.e. the guaranteed convergence rate ψ of MAS with different self-delay is never larger than for MAS without self-delay or with identical self-delay. This is not surprising since MAS with different self-delay contain the other two configurations as special cases. The following section compares the conditions of Theorem 3.17 to 3.19 in more detail.

3.8.2 Maximal Guaranteed Convergence Rate

In many single integrator MAS applications, we may tune the gain K in order to maximize the convergence rate ψ for unknown topologies and delays. An important question is therefore: how do we choose K in order to maximize ψ for a given lower bound on the connectivity $\overline{\lambda}_2$ but arbitrary bounded delays $\tau_{ij}, T_{ij} \leq \mathcal{T}$? This optimization problem is solved in this subsection based on (3.46), (3.47), and (3.48), respectively.

First, we consider MAS with identical self-delay, i.e. (3.47). Remember that $K < \frac{1}{2\mathcal{T}}$. Consider (3.47) and note that the right hand side increases as K decreases. Starting from $K = \frac{1}{2\mathcal{T}}$ and decreasing K, (3.47) allows for larger ψ until $\psi = (\overline{\lambda}_2 - \varepsilon)K$ for a very small $\varepsilon > 0$. Thus, the optimal gain K_{opt} satisfies

$$2\mathcal{T}K_{opt}e^{(\overline{\lambda}_2-\varepsilon)K_{opt}\mathcal{T}} = 1,$$

with optimal convergence rate $\psi_{opt} = (\overline{\lambda}_2 - \varepsilon)K_{opt}$. A similar condition for MAS without self-delay and MAS with different self-delay is much more difficult to obtain because of the term $1 - \frac{\psi}{K} + \overline{\lambda}_2$ in (3.46) and (3.48). This nonlinear optimization problem can be solved numerically in MATLAB® using `fmincon` with starting point $K = \frac{1}{\mathcal{T}}$ for MAS without self-delay and $K = \frac{1}{2\mathcal{T}}$ for MAS with different self-delay, respectively. The optimization results for all three cases are presented in Figure 3.11.

As in the undelayed case, the convergence rate increases with increasing minimal connectivity $\overline{\lambda}_2$, see Figure 3.11(a). Moreover, we see that increasing the delay bound $\mathcal{T}$ decreases the convergence rate. This underlines our previous statement: consensus in single integrator MAS without self-delay is reached independent of delay; yet, a fast convergence can only be guaranteed for small delays and appropriately chosen K. Figure 3.11(a) also shows that the guaranteed convergence rate ψ for MAS without self-delay is greater than for MAS with identical self-delay in all studied cases. This difference is also due to different optimal gains K_{opt} for both cases, see Figure 3.11(b). In this example, the optimal gains for MAS without self-delay are roughly twice the optimal gains of MAS with identical or different self-delays for identical $\overline{\lambda}_2$ and $\mathcal{T}$.

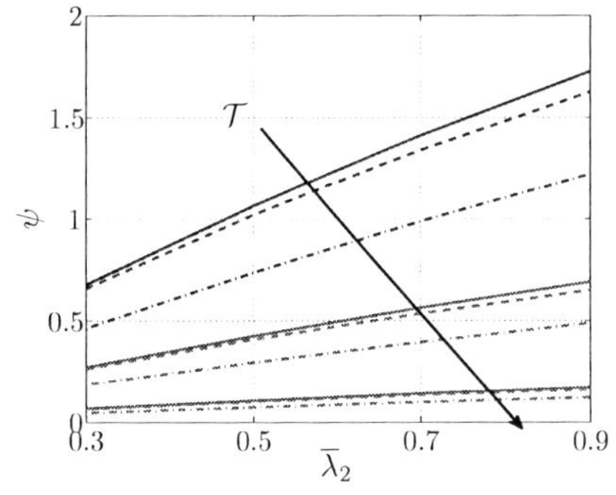

(a) Convergence rate ψ vs. connectivity bound $\overline{\lambda}_2$.

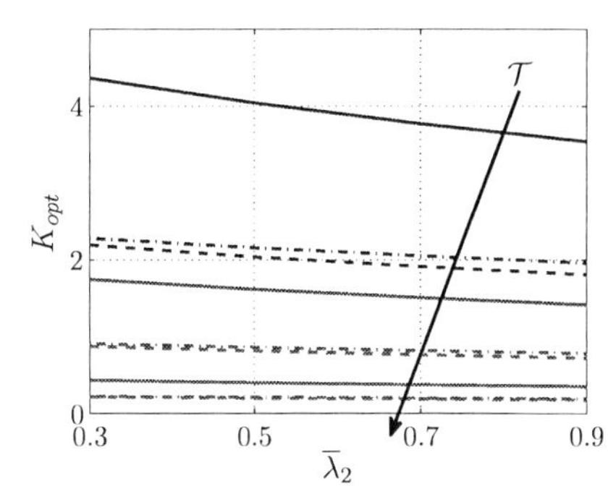

(b) Optimal gain K_{opt} vs. connectivity bound $\overline{\lambda}_2$.

Figure 3.11: Maximal convergence rate ψ and optimal gain K_{opt} vs. connectivity bound $\overline{\lambda}_2$ for MAS without self-delay (solid lines), MAS with identical self-delay (dashed lines), and MAS with different self-delay (dash-dotted lines) for different delay bounds $\mathcal{T} = 0.2$ (blue), $\mathcal{T} = 0.5$ (green), and $\mathcal{T} = 2$ (red). The arrow indicates how ψ and K_{opt} decrease as $\mathcal{T}$ increases.

3.8.3 Example: Maximal Guaranteed Convergence Rate

We illustrate the accuracy and conservatism of this maximal guaranteed convergence rate property in the following example: Consider a MAS without self-delay consisting of four single integrator agents (3.35). These agents are interconnected on an undirected, connected graph with unknown topology. Yet, we know that all edges have weight 1 and straightforward combinatorics show that there are six possible topologies with different spectrum of the corresponding Laplacian matrices. The smallest second smallest eigenvalue of $\overline{L}$ of all these topologies is obtained for an undirected line graph, i.e.

$$D_1^{-1}A_1 = \begin{bmatrix} 0 & 1 & 0 & 0 \\ 0.5 & 0 & 0.5 & 0 \\ 0 & 0.5 & 0 & 0.5 \\ 0 & 0 & 1 & 0 \end{bmatrix}, \tag{3.49}$$

with spectrum $\sigma(D_1^{-1}A_1) = \{-1, -0.5, 0.5, 1\}$. Hence, we have $\overline{\lambda}_2 = 0.5$ for all possible undirected, connected topologies. Moreover, we assume that all delays are bounded by $\mathcal{T} = 0.5$.

Using the results of Section 3.8.2, we obtain a maximal guaranteed convergence rate of $\psi = 0.4253$ if we choose $K = 1.6169$ for given $\overline{\lambda}_2 = 0.5$ and $\mathcal{T} = 0.5$. We compare this result with a numerical computation of the rightmost roots of the closed loop MAS with traceDDE (Breda et al., 2010). For the MAS with line graph topology (3.49), $K = 1.6169$, and identical delays $\tau_{ij} = \mathcal{T} = 0.5$ for all i, j, the rightmost nonzero root of $\Delta(s)$ is located at -0.55. Hence, the convergence rate of this MAS is only slightly larger than the guaranteed convergence rate $\psi = 0.4253$.

We compare this result to a MAS where the agents are all-to-all connected, i.e.

$$D_2^{-1}A_2 = \begin{bmatrix} 0 & \frac{1}{3} & \frac{1}{3} & \frac{1}{3} \\ \frac{1}{3} & 0 & \frac{1}{3} & \frac{1}{3} \\ \frac{1}{3} & \frac{1}{3} & 0 & \frac{1}{3} \\ \frac{1}{3} & \frac{1}{3} & \frac{1}{3} & 0 \end{bmatrix}, \tag{3.50}$$

with spectrum $\sigma(D_2^{-1}A_2) = \{-\frac{1}{3}, -\frac{1}{3}, -\frac{1}{3}, 1\}$. We consider again identical delays $\tau_{ij} = \mathcal{T} = 0.5$ for all i, j. In this case, numerical computations indicated the rightmost nonzero root of $\Delta(s)$ at $-2.95 \pm j1.94$, i.e. the MAS converges much faster than the guaranteed convergence rate $\psi = 0.4253$.

This example nicely illustrates that the convergence rate conditions in Theorems 3.17, 3.18, and 3.19 may be accurate or conservative, depending on the topology between the agents. This is not surprising as these theorems provide a *robust* convergence rate for arbitrary topologies and delays. This robustness is a big advantage compared to, e.g., a numerical computation of the convergence rate. This numerical computation requires the exact knowledge of the topology and the delays. Yet, in many MAS, neither the exact topology nor the exact delays are known.

Summarizing, we see that the conditions in Theorem 3.17, 3.18, and 3.19 are useful for gain optimization in order to increase the convergence rate. We also found out that MAS without self-delays provide a higher guaranteed convergence rate than MAS with self-delays in all studied cases.

3.9 Summary

In this chapter, we presented a method to analyse the delay robustness of consensus and flocking in linear MAS on undirected graphs. Thereby, we compared three different feedback delay configurations, i.e. without self-delay, with identical self-delay, and with different self-delay. It is a big advantage of the developed method that all these delay configurations are investigated in the same framework. Moreover, the main result in Theorem 3.6 states a unifying condition for these three cases. This enables an easy comparison of the delay robustness of these delay configurations. It turned out that MAS without self-delay are in many important cases more robust to delays than MAS with self-delays. In addition, we found out that, in general, larger delays require smaller gains. Finally, the developed method has been applied to many standard MAS problems, such as consensus in single integrator MAS, and most interestingly, we derived robust and scalable convergence rate conditions for single integrator MAS.

Chapter 4

Nonlinear Multi-Agent Systems with Relative Degree One

In the previous chapter, delay robustness in linear MAS has been investigated. Now, we consider nonlinear agent dynamics with relative degree one interconnected on directed, switching graphs as described in Section 4.1. We show in Section 4.2 that this MAS achieves rendezvous for arbitrarily large delays if the coupling functions are such that the agent dynamics show an integrating behavior. This holds for all three delay models: constant, time-varying, and distributed delays. We show how this integrating behavior can be achieved in special system classes, like input affine systems in Section 4.3. Finally, we use the developed conditions to investigate the synchronization of Kuramoto oscillators with communication delays in Section 4.4. A summary of the results of this chapter is given in Section 4.5. Preliminary results of this chapter have been published in Münz et al. (2009b,d, 2008a, 2007b).

4.1 Multi-Agent Systems with Relative Degree One

4.1.1 Agent Dynamics

We consider nonlinear relative degree one MAS, for short *nonlinear RD1 MAS*, with agent dynamics

$$\Sigma_i : \begin{cases} \dot{x}_i(t) &= f_i(x_i(t), u_i(t)) \\ y_i(t) &= h_i(x_i(t)) \end{cases}, \qquad i \in \mathcal{N} = \{1, \ldots, N\}, \tag{4.1}$$

where $x_i(t) \in \mathbb{R}^{n_i}, u_i(t) \in \mathbb{R}^m, y_i(t) \in \mathbb{R}^m$ are the state, input, and output of agent i, respectively. The functions $f_i : \mathbb{R}^{n_i} \times \mathbb{R}^m \to \mathbb{R}^{n_i}$ and $h_i : \mathbb{R}^{n_i} \to \mathbb{R}^m$ are Lipschitz continuous. For illustration, we may think of a MAS consisting of robots, where y_i and u_i are the velocity output and force input or the position output and velocity input of a robot. The initial condition of (4.1) interconnected with delays as defined below is $\varphi \in \mathcal{C}_{\sum n_i} = \mathcal{C}([-\mathcal{T}, 0], \mathbb{R}^{\sum n_i})$. Moreover, we assume the following:

Assumption 4.1. *The agents* (4.1) *have uniform relative degree one, i.e.* $\frac{\partial \dot{y}_i}{\partial u_i}(x_i, u_i) = \frac{\partial h_i}{\partial x_i}\frac{\partial f_i}{\partial u_i}(x_i, u_i)$ *is non-singular for all* $x_i \in \mathbb{R}^{n_i}, u_i \in \mathbb{R}^m$.

For simplicity of presentation, we often say relative degree when referring to uniform relative degree, see Sepulchre et al. (1997) for details. Remember that nonlinear systems with relative degree one is a large class of systems, including for example passive systems or Euler-Lagrange systems. In fact, most physical systems are passive if the output is properly chosen. Yet, this passive output sometimes does not coincide with the measurable output. Therefore, Assumption 4.1 can also be seen as a condition on the measurable output, e.g. it may require velocity measurements instead of position measurements. This issue will be discussed in more detail in Section 5.1.

4.1.2 Interconnection with Delays

The aim of the nonlinear RD1 MAS is to achieve a rendezvous in the output, i.e.

$$\begin{aligned} \lim_{t\to\infty} \|y_i(t) - y_j(t)\| &= 0, \qquad \text{for all } i,j \in \mathcal{N}, \\ \lim_{t\to\infty} \dot{y}_i(t) &= 0, \qquad \text{for all } i \in \mathcal{N}. \end{aligned} \tag{4.2}$$

In order to achieve this cooperation, the agents are interconnected in a network with possibly changing topology, i.e. the underlying graph may be switching. We consider the following two feedback interconnections

$$u_i(t) = -\sum_{j=1}^{N} \frac{a_{ij}(t)}{d_i(t)} \tilde{k}_{ij}\left(x_i(t), y_i(t) - \mathbb{T}_{ij}(y_{j,t})\right), \tag{4.3}$$

$$u_i(t) = -\sum_{j=1}^{N} \frac{a_{ij}(t)}{d_i(t)} k_{ij}\left(y_i(t) - \mathbb{T}_{ij}(y_{j,t})\right), \tag{4.4}$$

where $\tilde{k}_{ij} : \mathbb{R}^{n_i} \times \mathbb{R}^m \to \mathbb{R}^m$, $k_{ij} : \mathbb{R}^m \to \mathbb{R}^m$ are nonlinear, Lipschitz continuous coupling functions that depend on the delayed output difference $y_i(t) - \mathbb{T}_{ij}(y_{j,t})$ between the agents i and j and, in the more general case (4.3), also on the state x_i of agent i. The functions $f_i, h_i, \tilde{k}_{ij}$, and k_{ij} are sufficiently smooth such that the closed loop system has piecewise continuously differentiable solutions, even for piecewise continuous time-varying delays, see Hale and Lunel (1993).

The elements of the adjacency matrix $a_{ij}(t)$ and the in-degree of agent i $d_i(t) = \sum_{j=1}^{N} a_{ij}(t)$ are piecewise constant because the underlying graph is switching. Switching graphs are particularly interesting because they can be used to model important network properties like packet-loss or limited communication range in wireless networks. In this chapter, we consider a switching graph $\mathcal{G} : \mathbb{R} \to \mathfrak{G}$ where $\mathfrak{G} = \{\mathcal{G}_p\}, p \in \mathcal{P} = \{1, \ldots, P\}$, is a finite set of P different directed graphs $\mathcal{G}_p = (\mathcal{V}, \mathcal{E}_p)$ with identical node set $\mathcal{V}$ but different edge set $\mathcal{E}_p$ and corresponding adjacency matrix $A_p \in \mathbb{R}^{N\times N}$. The edge set $\mathcal{E}(t) = \mathcal{E}_p$ and the adjacency matrix $A(t) = A_p$ at time t correspond to the graph $\mathcal{G}(t) = \mathcal{G}_p$. All graphs $\mathcal{G}_p$ are directed, i.e. $a_{ij}(t) = a_{ji}(t)$ is not required. The function $\mathcal{G}$ is piecewise constant from the right and we denote the time instances where $\mathcal{G}$ switches $t_\varsigma > t_{\varsigma-1}, \varsigma = 1, 2, \ldots$. We assume there are infinitely many such switching times because otherwise we could just analyse the last active graph. Moreover, any two consecutive switching instants are separated by a dwell-time h_{DW}, i.e. $t_\varsigma - t_{\varsigma-1} \geq h_{DW}$. This guarantees that the switching graph is non-chattering. We assume that the graph is uniformly quasi-strongly connected which is defined next. For this definition, we use the following notation for a union graph $\mathcal{G}([t_1, t_2]) = (\mathcal{V}, \bigcup_{t\in[t_1,t_2]} \mathcal{E}(t))$ over the interval $[t_1, t_2]$. A union graph consists of all vertices in $\mathcal{V}$ and all edges that appear at any time $t \in [t_1, t_2]$.

Definition 4.2 (Uniformly quasi-strongly connected). *A switching graph $\mathcal{G} : \mathbb{R} \to \mathfrak{G}$ is* uniformly quasi-strongly connected *if there exists a $\mathfrak{T} > 0$ such that, for all $t \geq 0$, the union graph $\mathcal{G}([t, t + \mathfrak{T}])$ is quasi-strongly connected.*

A graph is uniformly quasi-strongly connected if the union graph over the interval $[t, t + \mathfrak{T}]$, for any t, is quasi-strongly connected, i.e. it contains a spanning tree. This is a very weak assumption on the connectivity. It may hold even if none of the graphs $\mathcal{G}_p \in \mathfrak{G}$ is quasi-strongly connected, even if every graph $\mathcal{G}_p$ has just one edge. Uniform quasi-strong connectivity is in fact the weakest assumption on the graph connectivity such that consensus can be proven. Counter examples are provided in Moreau (2005); Lin (2006) which show that, in general, consensus is not achieved in single integrator MAS if the graph is not uniformly quasi-strongly connected, e.g. if $\mathfrak{T}$ in Definition 4.2 is not a constant.

In the previous chapter, we studied the cooperative behavior for different feedback delay configurations, i.e. with or without self-delay. In the remaining two chapters, we concentrate on MAS without self-delay for two reasons:

- Chapter 3 revealed that MAS without self-delay usually more robust to delays than MAS with self-delay, see Section 3.4.3 for example.
- MAS without self-delay are a suitable model for communication and coupling delays, see Section 2.6, which is the main motivation for this thesis.

In contrast to Chapter 3, we consider now all three delay models for the delay operator $\mathbb{T}_{ij}(y_{j,t})$, i.e. constant, time-varying, or distributed delays, see Section 2.6. In all cases, we require that the delays are bounded, i.e. there exists a finite $\mathcal{T} > 0$ such that $\tau_{ij}(t) \leq \mathcal{T}$ for constant and time-varying delays and $\phi_{ij}(\eta) = 0$ for $\eta > \mathcal{T}$ for the kernel of the distributed delays, see Equation (2.16). The time-varying delays are piecewise continuous, see Section 2.6. In Section 3.3, we showed that linear MAS without self-delay can only achieve static consensus. Using similar arguments in the time domain, we can show that MAS (4.1) with feedback (4.3) in general achieves rendezvous but not flocking, see Section 2.5.

Finally, we assume that the coupling functions $\tilde{k}_{ij}, k_{ij}$ satisfy the following design condition:

Design Condition 4.3. *The coupling functions $\tilde{k}_{ij} : \mathbb{R}^{n_i} \times \mathbb{R}^m \to \mathbb{R}^m$ are such that all outputs $y_i(t)$ of the closed loop system (4.1), (4.3) satisfy*

$$\dot{y}_i(t) = \sum_{j\in\mathcal{N}_i(t)} \widehat{k_{ij}}(t) \left(\mathbb{T}_{ij}(y_{j,t}) - y_i(t)\right), \qquad \textit{for all } t \geq 0, \tag{4.5}$$

for any $\widehat{k_{ij}} : \mathbb{R}_0^+ \to \mathbb{R}_0^+$ such that the following holds:

- *there exists a* $\overline{\mathcal{K}} > 0$*, such that* $\sum_{j \in \mathcal{N}_i(t)} \widehat{k_{ij}}(t) \leq \overline{\mathcal{K}}$ *for all* $i \in \mathcal{N}$ *and* $t \geq 0$ *and*
- *there exists a continuous, non-decreasing function* $\underline{\mathcal{K}} : \mathbb{R}_0^+ \to \mathbb{R}_0^+$ *with* $\underline{\mathcal{K}}(\eta) > 0$ *for* $\eta > 0$ *such that* $\widehat{k_{ij}}(t) \geq \underline{\mathcal{K}}(\eta)$ *for all* $i \in \mathcal{N}, j \in \mathcal{N}_i(t)$*, and all* $t \geq 0$ *where* $\|\mathbb{T}_{ij}(y_{j,t}) - y_i(t)\| \geq \eta$.

Design Condition 4.3 assumes an *integrating behavior*, i.e. the output $y_i(t)$ of each agent integrates over a weighted sum of the delayed output differences $\mathbb{T}_{ij}(y_{j,t}) - y_i(t)$ to all parents. If the agents (4.1) are single integrators, then Design Condition 4.3 is satisfied for very general class of passive coupling functions k_{ij}, see Corollary 4.13 below. In fact, it is quite natural to assume that agents, which aim for rendezvous, show an integrating behavior from input to output. This integrating behavior implies that the output remains constant for zero inputs and the main difficulty is to find coupling functions such that all outputs eventually remain constant at the same point. On the other hand, the integrating behavior in (4.5) implies that each agents tracks the delayed outputs of its parents. In other words, the output of each agent varies toward the outputs of its parents, or, more precisely, some weighted average of the last known outputs of its parents. The parameter $\overline{\mathcal{K}}$ corresponds to an upper bound on how fast agent i moves depending on the output differences. The function $\underline{\mathcal{K}}$ guarantees that agent i takes all its neighbours into account instead of tracking some neighbours, completely ignoring others. Note that $\overline{\mathcal{K}}$ and $\underline{\mathcal{K}}$ are upper and lower bound on $\widehat{k_{ij}}(t)$ for all $t \geq 0$, i.e. they depend on the initial conditions, and in particular they may increase with increasing initial conditions.

If the agents themselves do not show an integrating behavior, then the coupling functions (4.3) have to include an appropriate input-output-transformation that achieves an integrating behavior as in (4.5). For input affine systems, this inner controller can be obtained by an input output linearization for example, see Corollary 4.9.

In Lin (2006), the author considers nonlinear MAS without delays

$$\dot{y}_i(t) = f_{i,p}(y_1, y_2, \dots, y_N), \tag{4.6}$$

where $p \in \mathcal{P}$ denotes the active graph. The main result states that this MAS achieves rendezvous on uniformly quasi-strongly connected graphs if the dynamics satisfy

$$f_{i,p}(y) \in \operatorname{ri}\left(\operatorname{TC}\left(y_i, \operatorname{Co}\left\{y_i, y_j : j \in \mathcal{N}_{i,p}\right\}\right)\right), \tag{4.7}$$

for all y, where $\operatorname{Co}\{\zeta_1, \dots, \zeta_n\}$ denotes the *convex hull* of $\{\zeta_1, \dots, \zeta_n\}$, $\operatorname{TC}(\zeta_1, \mathfrak{M})$ is the tangent cone to the set $\mathfrak{M}$ at the point ζ_1, and $\operatorname{ri}(\cdot)$ describes the *relative interior* of a set, see Lin (2006); Boyd and Vandenberghe (2004). Condition (4.7) is equivalent to

$$\dot{y}_i(t) = \sum_{j \in \mathcal{N}_i(t)} \widehat{k_{ij}}(t)\left(y_j(t) - y_i(t)\right), \qquad \text{for all } i \in \mathcal{N}, t \geq 0, \tag{4.8}$$

where the functions $\widehat{k_{ij}} : \mathbb{R} \to \mathbb{R}$ satisfy $\widehat{k_{ij}}(t) > 0$ for all $i, j \in \mathcal{N}$. Condition (4.5) is in fact an extension of this condition in Lin (2006) for nonlinear agents with internal dynamics. For agent dynamics (4.6), there exists an upper bound $\overline{\mathcal{K}}$ on $\sum_{j\in\mathcal{N}_i(t)} \widehat{k_{ij}}(t)$ as requires in Design Condition 4.3 because the solutions of (4.6) are bounded, see also the proof of Corollary 4.9 below or the proof in Lin (2006). The time-invariant lower bound $\underline{\mathcal{K}}$ can be derived from (4.6),(4.7) because Lin (2006) considers dynamics $f_{i,p}(y)$ without internal dynamics, see also the proof of Corollary 4.9. As we consider here the more general case with internal dynamics, we assume a time-invariant lower bound on the influence of agent j on agent i if $j \in \mathcal{N}_i(t)$. A very similar assumption on the minimal influence appears as δ-connected graphs in Moreau (2004), where the author considers linear MAS without delays. For a MAS without internal dynamics, we will relax Design Condition 4.3 in Corollary 4.9. In summary, Design Condition 4.3 generalized previous conditions on MAS without delays.

We show first in Section 4.2 that nonlinear RD1 MAS (4.1) with feedback interconnections (4.3) and (4.4) satisfying Design Condition 4.3 achieves rendezvous on switching graphs. The case of constant topologies is treated as special case. We derive equivalent conditions for special classes of MAS with dynamics (4.1) in Section 4.3. Finally, we illustrate the relevance of these results investigating the synchronization of Kuramoto oscillators in Section 4.4.

4.2 Rendezvous on Constant and Switching Topologies

In this section, we present the main result of this chapter in Theorem 4.5. It states that rendezvous is guaranteed in nonlinear RD1 MAS (4.1) with suitable coupling functions $\tilde{k}_{ij}, k_{ij}$ satisfying Design Condition 4.3.

The proof is based on a contraction argument that was used in Moreau (2004) for linear MAS and in Lin (2006) for nonlinear MAS without delays. We extend this result to nonlinear RD1 MAS with heterogeneous delays. The fundamental idea of the proof is to define a time-varying convex set $\Upsilon : \mathbb{R} \to \mathbb{R}^m$, such that $\Upsilon(t)$ contains all output trajectory pieces $y_{i,t}$, i.e. all $y_i(t+\eta), \eta \in [-\mathcal{T}, 0]$. Then, we show by an iterative procedure that this set contracts to a single point as $t \to \infty$.

The set $\Upsilon(t) \subset \mathbb{R}^m$ is defined as the smallest hyper-rectangle, such that the surfaces of Υ are aligned with the coordinate axes and $\Upsilon(t)$ contains all output trajectory pieces $y_{i,t}$, see Figure 4.1 for an exemplary illustration. The boundary of $\Upsilon(t)$ is denoted $\partial\Upsilon(t)$. The $2m$ surfaces of $\Upsilon(t)$ are $\overline{\Upsilon_k}(t)$ and $\underline{\Upsilon_k}(t)$, $k = 1, \dots, m$, such that $\overline{\Upsilon_k}(t) \geq y_{i[k]}(t+\eta) \geq \underline{\Upsilon_k}(t)$ for all $i \in \mathcal{N}$ and for all $\eta \in [-\mathcal{T}, 0]$, where $y_{i[k]}(t+\eta)$ denotes the k-th element of $y_i(t+\eta)$. Moreover, we define the length of the hyper-rectangle in dimension k as $\widehat{\Upsilon_k}(t) = \overline{\Upsilon_k}(t) - \underline{\Upsilon_k}(t)$. Since $\Upsilon(t)$ is the smallest hyper-rectangle, there is always at least one $i \in \mathcal{N}$ such that $y_{i[k]}(t+\eta) = \overline{\Upsilon_k}(t)$ for some $\eta \in [-\mathcal{T}, 0]$ and similarly at least one $j \in \mathcal{N}$ such that $y_{j[k]}(t+\eta) = \underline{\Upsilon_k}(t)$ for some $\eta \in [-\mathcal{T}, 0]$, see Figure 4.1. Clearly, the hyper-rectangle $\Upsilon(t)$ changes over time. Since we consider rendezvous of the outputs and not rendezvous of the states, it suffices to consider the output trajectories but not the state trajectories. Note however that implicit conditions on the states of the agents are imposed by Condition (4.5), which has to hold for all $t \geq 0$.

In the following lemma, we show that the set $\Upsilon(t)$ is non-increasing, i.e. $\Upsilon(t_2) \subseteq \Upsilon(t_1)$

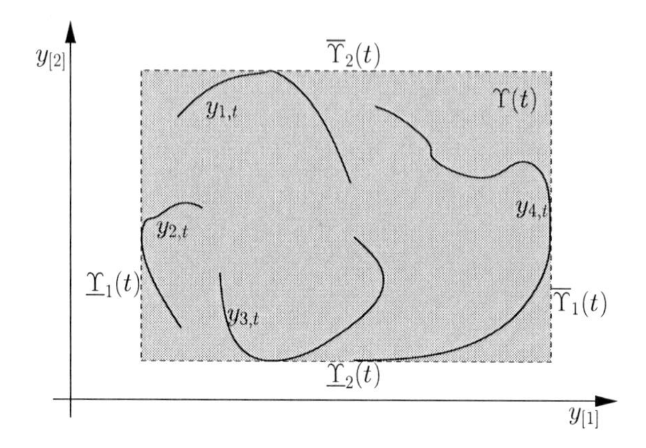

Figure 4.1: Exemplary illustration of hyper-rectangle $\Upsilon(t)$ for a MAS with four agents and $y_i(t) \in \mathbb{R}^2$.

for all $t_2 \geq t_1$. This implies that $\Upsilon(0)$ is positively invariant.

Lemma 4.4. *The set $\Upsilon(t)$ defined above is non-increasing for MAS on networks of arbitrary size $N \in \mathbb{N}$ with arbitrary bounded heterogeneous constant, time-varying, or distributed delays if the agent dynamics* (4.1) *satisfy Assumption 4.1 and the continuous coupling functions $\tilde{k}_{ij} : \mathbb{R}^{n_i} \times \mathbb{R}^m \to \mathbb{R}^m$ are such that all outputs $y_i(t)$ of the closed loop system* (4.1), (4.3) *satisfy*

$$\dot{y}_i(t) = \sum_{j \in \mathcal{N}_i(t)} \widehat{k_{ij}}(t) \left(\mathbb{T}_{ij}(y_{j,t}) - y_i(t) \right), \qquad \text{for all } t \geq 0, \tag{4.9}$$

for any $\widehat{k_{ij}} : \mathbb{R}_0^+ \to \mathbb{R}_0^+$, i.e. $\widehat{k_{ij}}(t) \geq 0$ for all t.

Note that Condition (4.9) is weaker than Condition (4.5) because $\widehat{k_{ij}}(t)$ may be zero.

Proof. The trajectories $y_i(t)$ are piecewise continuously differentiable because the system has relative degree one and the dynamics as well as the coupling functions are sufficiently smooth. The set $\Upsilon(t)$ can only increase at time t if there is an agent i on the boundary of Υ, i.e. with $y_i(t) \in \partial\Upsilon(t)$, and if this agent is leaving $\Upsilon(t)$. We show briefly that any $y_i(t) \in \partial\Upsilon(t)$ is not leaving $\Upsilon(t)$. Condition (4.9) implies that $\dot{y}_i(t)$ is either zero or heading toward some weighted average of the past outputs of its parents $\mathbb{T}_{ij}(y_{j,t}), j \in \mathcal{N}_i(t)$. Remember that all trajectory pieces $y_{j,t}$ are in $\Upsilon(t)$, see Figure 4.1. Therefore, $\mathbb{T}_{ij}(y_{j,t}) \in \Upsilon(t)$ for constant, time-varying, and distributed delays. This also holds for distributed delays because $\int_0^{\mathcal{T}_{ij}} \phi_{ij}(\eta) y_j(t-\eta) d\eta \in \mathrm{Co}\{y_j(t+\eta) : \eta \in [-\mathcal{T}, 0]\} \subseteq \Upsilon(t)$ for any non-negative delay kernel ϕ_{ij} with $\int_0^{\mathcal{T}_{ij}} \phi_{ij}(\eta) d\eta = 1$, see (3.22). Thus, the derivative

$\dot{y}_i(t)$ of an agent with $y_i(t) \in \partial\Upsilon(t)$ is such that the output of the agent either remains on the boundary or varies toward the interior of $\Upsilon(t)$. Summarizing, y_i is not leaving $\Upsilon(t)$ and $\Upsilon(t)$ is non-increasing. □

Now, we state the main result of this chapter:

Theorem 4.5 (Rendezvous in nonlinear RD1 MAS with switching topology). *A MAS with agent dynamics* (4.1) *that satisfy Assumption 4.1 and with interconnection* (4.3) *where $\tilde{k}_{ij}$ satisfy Design Condition 4.3 achieves rendezvous asymptotically on networks of arbitrary size $N \in \mathbb{N}$, with arbitrary uniformly quasi-strongly connected switching directed topologies, and for arbitrary bounded heterogeneous constant, time-varying, or distributed delays.*

Proof. We have shown in Lemma 4.4 that $\Upsilon(t)$ is non-increasing. It remains to show that $\lim_{t\to\infty} \Upsilon(t) = \{y^*\}$ for some $y^* \in \mathbb{R}^m$, i.e. the hyper-rectangle contracts to a single point. This is equivalent to proving that the length of the hyper-rectangle in every direction shrinks to zero, i.e. $\lim_{t\to\infty} \widehat{\Upsilon_k}(t) = 0$ for all $k = 1, \ldots, m$. In the sequel, we consider only one dimension, i.e. one arbitrary k, and show that $\lim_{t\to\infty} \widehat{\Upsilon_k}(t) = 0$. The same argument applies for all other dimensions.

Remember that all delays are bounded by $\mathcal{T} \in \mathbb{R}$ and the graph is uniformly quasi-strongly connected, i.e. there exists a $\mathfrak{T} > 0$ such that the union graph $\mathcal{G}([t, t+\mathfrak{T}])$ is quasi-strongly connected for any $t \geq 0$. Given any time $t_0 \geq 0$, we show in the sequel that

$$\widehat{\Upsilon_k}(t_0 + T_{\text{iter}}) - \widehat{\Upsilon_k}(t_0) \leq -\varepsilon(\widehat{\Upsilon_k}(t_0)), \tag{4.10}$$

where $T_{\text{iter}} = 2N(\mathcal{T} + \mathfrak{T}) + \mathcal{T}$, N is the number of agents, and $\varepsilon : \mathbb{R}_0^+ \to \mathbb{R}_0^+$ is a non-decreasing function with $\varepsilon(\eta) > 0$ for $\eta > 0$. Using Equation (4.10), we show in the last step that $\lim_{t\to\infty} \widehat{\Upsilon_k}(t) = 0$.

In order to establish (4.10), we develop an iterative procedure that shows that all agents that are close to one of the surfaces, either $\overline{\Upsilon_k}(t_0)$ or $\underline{\Upsilon_k}(t_0)$, at time t_0 have at least distance $\varepsilon(\widehat{\Upsilon_k}(t_0))$ to this surface at time $t_0 + T_{\text{iter}}$. There are two key elements of the proof corresponding to $\overline{\mathcal{K}}, \underline{\mathcal{K}}$ in Design Condition 4.3: (i) the upper bound $\overline{\mathcal{K}}$ guarantees that agents in the interior of $\Upsilon(t)$ approach the boundary $\partial\Upsilon(t)$ at most exponentially and (ii) the lower bound $\underline{\mathcal{K}}$ implies that agents on the boundary eventually move to the interior of $\Upsilon(t)$ because they are at some point connected to other agents in the interior.

Overview of iterative procedure The arguments explained below are iterated over $2N$ consecutive intervals $\mathfrak{I}_l = [t_0 + (l-1)(\mathcal{T} + \mathfrak{T}), t_0 + l(\mathcal{T} + \mathfrak{T})]$, where the intervals and the corresponding iteration steps are indexed by $l \in \mathfrak{N} = \{1, \ldots, 2N\}$. The particular structure of these intervals is explained at the end of the proof. Figure 4.2 exemplarily shows the first two intervals $\mathfrak{I}_1, \mathfrak{I}_2$. In each iteration interval, there is a subinterval $\hat{\mathfrak{I}}_l = [t_0 + l\mathcal{T} + (l-1)\mathfrak{T}, t_0 + l(\mathcal{T} + \mathfrak{T})] \subset \mathfrak{I}_l$ of length $\mathfrak{T}$ for every $l \in \mathfrak{N}$, see $\hat{\mathfrak{I}}_1, \hat{\mathfrak{I}}_2$ in Figure 4.2. Since the graph is uniformly quasi-strongly connected, the union graph $\mathcal{G}(\hat{\mathfrak{I}}_l)$ has at least one root. We denote the index of this root $i_{R_l}, l \in \mathfrak{N}$. If there are several

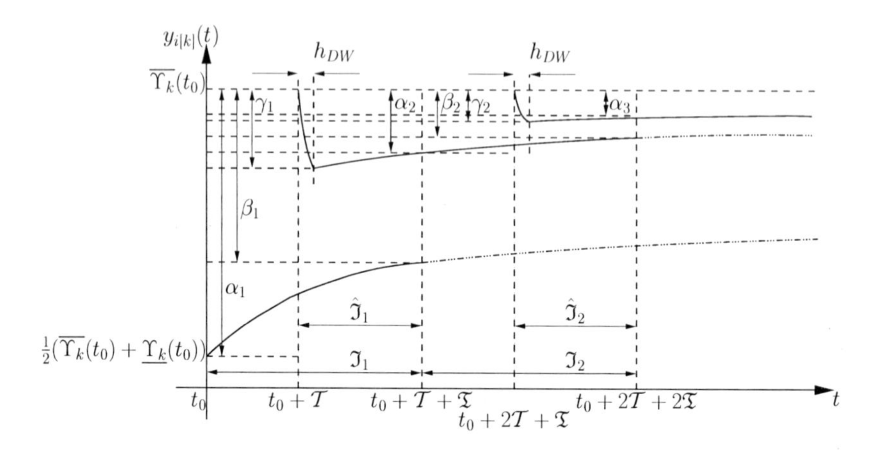

Figure 4.2: Schematic illustration of two iterations of the contraction argument.

roots, we choose any one of them. Note that some of these roots $i_{R_l}, l \in \mathfrak{N}$, correspond to the same node because the MAS has N agents and there are $2N$ intervals $\hat{\mathfrak{I}}_l$, i.e. there will always be $l, \tilde{l} \in \mathfrak{N}$ with $l \neq \tilde{l}$ such that $i_{R_l} = i_{R_{\tilde{l}}}$. Now, consider the k-th element of the output vectors $y_{i_{R_l}[k]}(t_0)$ for $l \in \mathfrak{N}$, i.e. the output at time t_0 of those agents that will become root in one of the $2N$ intervals $\hat{\mathfrak{I}}_l$. This output $y_{i_{R_l}[k]}(t_0)$ of at least N of all $l \in \mathfrak{N}$ is either in the upper or lower half, i.e. either $y_{i_{R_l}[k]}(t_0) \in (\overline{\Upsilon_k}(t_0) - \frac{1}{2}\widehat{\Upsilon_k}(t_0), \overline{\Upsilon_k}(t_0)]$ or $y_{i_{R_l}[k]}(t_0) \in [\underline{\Upsilon_k}(t_0), \overline{\Upsilon_k}(t_0) - \frac{1}{2}\widehat{\Upsilon_k}(t_0)]$. We assume without loss of generality that at least N of these roots are in the lower half $[\underline{\Upsilon_k}(t_0), \overline{\Upsilon_k}(t_0) - \frac{1}{2}\widehat{\Upsilon_k}(t_0)]$.

The iteration works as follows: At the beginning of each interval $\mathfrak{I}_l$, we know that there is a nonempty set $\mathcal{I}_l \subseteq \mathcal{N}$ of agents that satisfies $y_{i[k]}(t_0 + (l-1)(\mathcal{T} + \mathfrak{T})) \leq \overline{\Upsilon_k}(t_0) - \alpha_l$ for some $\alpha_l > 0$, i.e. all agents $i \in \mathcal{I}_l$ have distance α_l from $\overline{\Upsilon_k}(t_0)$ at the beginning of the interval, see α_1, α_2 in Figure 4.2. The value α_l is initialized with $\alpha_1 = \frac{1}{2}\widehat{\Upsilon_k}(t_0)$. Remember from the previous paragraph that $\mathcal{I}_1$ contains at least N of the $2N$ roots i_{R_l}. This guarantees that $\mathcal{I}_1$ is not empty. Using $\overline{\mathcal{K}}$, we determine in each step a $\beta_l \in (0, \alpha_l)$ such that $y_{i[k]}(t_0 + l(\mathcal{T} + \mathfrak{T}) + \eta) \leq \overline{\Upsilon_k}(t_0) - \beta_l$ for all $\eta \in [-\mathcal{T} - \mathfrak{T}, 0]$ and all $i \in \mathcal{I}_l$. That is, all agents $i \in \mathcal{I}_l$ have distance β_l from $\overline{\Upsilon_k}(t_0)$ during the whole interval $\mathfrak{I}_l$, see β_1, β_2 in Figure 4.2. We will show later on that $\alpha_{l+1} \leq \beta_l$ for all l. Therefore, we have $\mathcal{I}_l \subseteq \mathcal{I}_{l+1}$, i.e. new elements can be added to the set $\mathcal{I}_l$ in each iteration step but elements cannot be removed. Thus, $\mathcal{I}_l$ always contains at least N roots i_{R_l}. This implies that in at least N of the $2N$ intervals $\hat{\mathfrak{I}}_l$ there is a root $i_{R_l} \in \mathcal{I}_l$. Thus, at least one of the remaining agents, say $I \in \mathcal{N} \backslash \mathcal{I}_l$, is at some time $t \in \hat{\mathfrak{I}}_l$ connected to an agent $J \in \mathcal{I}_l$ because the graph is uniformly quasi-strongly connected. For these N iteration steps, we use $\underline{\mathcal{K}}$ to determine a $\alpha_{l+1} \in (0, \beta_l)$ such that $y_{i[k]}(t_0 + l(\mathcal{T} + \mathfrak{T})) \leq \overline{\Upsilon_k}(t_0) - \alpha_{l+1}$

for all $i \in \mathcal{I}_l \cup \{I\}$. In other words, all agents $i \in \mathcal{I}_l \cup \{I\}$ have distance α_{l+1} from $\overline{\Upsilon_k}(t_0)$ at the end of the interval $\mathfrak{I}_l$. In those iteration steps where such an I exists, the iteration step ends with $\mathcal{I}_{l+1} = \mathcal{I}_l \cup \{I\}$. In the remaining iteration steps, i.e. if no such I exists, the iteration ends with $\mathcal{I}_{l+1} = \mathcal{I}_l$ and $\alpha_{l+1} = \beta_l$. After $2N$ steps, we know that $\mathcal{I}_{2N+1} = \mathcal{N}$ and therefore all $i \in \mathcal{N}$ satisfy $y_{i[k]}(t_0 + 2N(\mathcal{T} + \mathfrak{T})) \leq \overline{\Upsilon_k}(t_0) - \alpha_{2N+1}$ for some $\alpha_{2N+1} > 0$. That is, all agents have distance α_{2N+1} from the upper surface $\overline{\Upsilon_k}(t_0)$. Therefore, we can determine a function ε that satisfies (4.10).

Iteration step The first two iteration steps are illustrated schematically in Figure 4.2. At time t_0, at least N out of all $l \in \mathfrak{N}$ satisfy $y_{i_{R_l}[k]}(t_0) \leq \overline{\Upsilon_k}(t_0) - \alpha_1$ where $\alpha_1 = \frac{1}{2}\widehat{\Upsilon_k}(t_0)$. These N i_{R_l} correspond to at least one node and guarantee that $\mathcal{I}_1$ is nonempty. Thus, starting parameters α_1 and $\mathcal{I}_1$ of the iteration are initialized as explained above. The following paragraphs explain the iteration steps in detail.

At the beginning of the interval, we know that $y_{i[k]}(t_0 + (l-1)(\mathcal{T} + \mathfrak{T})) \leq \overline{\Upsilon_k}(t_0) - \alpha_l$ for all $i \in \mathcal{I}_l$. Condition (4.5) implies

$$\dot{y}_{i[k]}(t) \leq \overline{\mathcal{K}} \max_{j \in \mathcal{N}_i(t)} \left(\mathbb{T}_{ij}(y_{j,t[k]}) - y_{i[k]}(t) \right),$$

i.e. all agents $i \in \mathcal{I}_l$ are attracted at most with convergence rate $\overline{\mathcal{K}}$ toward the upper surface $\overline{\Upsilon_k}(t_0)$ because $\mathbb{T}_{ij}(y_{j,t[k]}) \leq \overline{\Upsilon_k}(t_0)$ for all $t \geq t_0$ and for all $j \in \mathcal{N}$. Hence, all agents $i \in \mathcal{I}_l$ satisfy $y_{i[k]}(t_0 + l(\mathcal{T} + \mathfrak{T}) + \eta) \leq \overline{\Upsilon_k}(t_0) - \beta_l$ for all $\eta \in [-(\mathcal{T} + \mathfrak{T}), 0]$, where $\beta_l = e^{-\overline{\mathcal{K}}(\mathcal{T}+\mathfrak{T})}\alpha_l$, see Figure 4.2.

Next, we consider the time-interval $\hat{\mathfrak{I}}_l = [t_0 + l\mathcal{T} + (l-1)\mathfrak{T}, t_0 + l(\mathcal{T} + \mathfrak{T})]$. The union graph $\mathcal{G}(\hat{\mathfrak{I}}_l)$ is quasi-strongly connected, i.e. there exists a spanning tree with root i_{R_l}. As explained above, we distinguish two cases: (i) $y_{i_{R_l}[k]}(t_0) > \overline{\Upsilon_k}(t_0) - \alpha_1$ and (ii) $y_{i_{R_l}[k]}(t_0) \leq \overline{\Upsilon_k}(t_0) - \alpha_1$. In the first case, the root i_{R_l} of this interval $\hat{\mathfrak{I}}_l$ was in the upper half at time t_0. Therefore, it is not clear if $i_{R_l} \in \mathcal{I}_l$ in this case. Thus, this iteration step ends with $\alpha_{l+1} = \beta_l = e^{-\overline{\mathcal{K}}(\mathcal{T}+\mathfrak{T})}\alpha_l$ and $\mathcal{I}_{l+1} = \mathcal{I}_l$.

In the second case, the root i_{R_l} of this interval $\hat{\mathfrak{I}}_l$ was in the lower half at time t_0, i.e. $i_{R_l} \in \mathcal{I}_1$. This case appears at least N times over all $2N$ iterations because at least N roots satisfy $i_{R_l} \in \mathcal{I}_1$, see above. Remember also that $\mathcal{I}_l \subseteq \mathcal{I}_{l+1}$, i.e. all roots $i_{R_l} \in \mathcal{I}_1$ also satisfy $i_{R_l} \in \mathcal{I}_l$ for all l. In this situation, the uniform quasi-strong connectivity implies that, as long as $\mathcal{I}_l \neq \mathcal{N}$, there is at least one agent $I \in \mathcal{N} \setminus \mathcal{I}_l$ that has a parent $J \in \mathcal{I}_l$ at some time $t \in \hat{\mathfrak{I}}_l$, i.e. $a_{IJ}(t) > 0$. This pair of agents I, J indeed exists because the root i_{R_l} is in $\mathcal{I}_l$. Since the graph switches with a dwell time h_{DW}, there exists a time interval $\Delta^l_{IJ} = [\underline{t}^l_{IJ}, \overline{t}^l_{IJ}] \subseteq \hat{\mathfrak{I}}_l$ with $\overline{t}^l_{IJ} - \underline{t}^l_{IJ} \geq h_{DW}$ such that $a_{IJ}(t) > 0$ for all $t \in \Delta^l_{IJ}$. During this interval, we know that $\mathbb{T}_{IJ}(y_{J,t[k]}) \leq \overline{\Upsilon_k}(t_0) - \beta_l$ because $J \in \mathcal{I}_l$. Now, consider the function $\underline{\mathcal{K}}$ and remember that $\underline{\mathcal{K}}(\eta) > 0$ for $\eta > 0$. There exists a $\underline{\mathcal{K}}_l > 0$ sufficiently small such that $\underline{\mathcal{K}}_l \leq \underline{\mathcal{K}}((1 - \frac{\underline{\mathcal{K}}_l}{\underline{\mathcal{K}}_l + \overline{\mathcal{K}}})\beta_l)$ because for $\underline{\mathcal{K}}_l \ll \overline{\mathcal{K}}$ we have $(1 - \frac{\underline{\mathcal{K}}_l}{\underline{\mathcal{K}}_l + \overline{\mathcal{K}}})\beta_l \approx \beta_l$ and $\underline{\mathcal{K}}(\beta_l) > 0$. Thus, we know that $\widehat{k_{IJ}}(t) \geq \underline{\mathcal{K}}_l$ as long as

$y_{I[k]}(t) - \mathbb{T}_{IJ}(y_{J,t[k]}) \geq (1 - \frac{\underline{\mathcal{K}}_l}{\underline{\mathcal{K}}_l + \overline{\mathcal{K}}})\beta_l$. Hence, Condition (4.5) implies that

$$\begin{aligned}\dot{y}_{I[k]}(t) &= \widehat{k_{IJ}}(t)\left(\mathbb{T}_{IJ}(y_{J,t[k]}) - y_{I[k]}(t)\right) + \sum_{j \in \mathcal{N}_I(t) \setminus \{J\}} \widehat{k_{Ij}}(t)\left(\mathbb{T}_{Ij}(y_{j,t[k]}) - y_{I[k]}(t)\right) \\ &\leq \underline{\mathcal{K}}_l\left(\overline{\Upsilon_k}(t_0) - \beta_l - y_{I[k]}(t)\right) + \overline{\mathcal{K}}\left(\overline{\Upsilon_k}(t_0) - y_{I[k]}(t)\right) \\ &= -(\underline{\mathcal{K}}_l + \overline{\mathcal{K}})(y_{I[k]}(t) - \overline{\Upsilon_k}(t_0)) - \underline{\mathcal{K}}_l\beta_l, \quad (4.11)\end{aligned}$$

as long as $y_{I[k]}(t) \geq \overline{\Upsilon_k}(t_0) - \frac{\underline{\mathcal{K}}_l}{\underline{\mathcal{K}}_l + \overline{\mathcal{K}}}\beta_l$. If this is the case on the whole interval Δ^l_{IJ}, we have $y_{I[k]}\left(\bar{t}^l_{IJ}\right) \leq \overline{\Upsilon_k}(t_0) - \gamma_l$ where $\gamma_l = (1 - e^{-(\underline{\mathcal{K}}_l + \overline{\mathcal{K}})h_{DW}})\frac{\underline{\mathcal{K}}_l}{\underline{\mathcal{K}}_l + \overline{\mathcal{K}}}\beta_l$, see Figure 4.2. If this is not the case at some time $t^* \in \Delta^l_{IJ}$, then we obviously have $y_{I[k]}(t^*) < \overline{\Upsilon_k}(t_0) - \frac{\underline{\mathcal{K}}_l}{\underline{\mathcal{K}}_l + \overline{\mathcal{K}}}\beta_l < \overline{\Upsilon_k}(t_0) - \gamma_l$. After this interval Δ^l_{IJ}, the topology might be such that agent I is attracted by a parent j with $\mathbb{T}_{ij}(y_{j,t[k]}) = \overline{\Upsilon_k}(t_0)$. The worst case is overestimated by assuming that I is attracted by such a parent with convergence rate $\overline{\mathcal{K}}$ during the complete interval $\hat{\mathfrak{I}}_l$. Thus, we obtain $y_{I[k]}(t_0 + l(\mathcal{T} + \mathfrak{T})) \leq \overline{\Upsilon_k}(t_0) - \alpha_{l+1}$ with

$$\alpha_{l+1} = e^{-\overline{\mathcal{K}}\mathfrak{T}}\gamma_l = (1 - e^{-(\underline{\mathcal{K}}_l + \overline{\mathcal{K}})h_{DW}})\frac{\underline{\mathcal{K}}_l}{\underline{\mathcal{K}}_l + \overline{\mathcal{K}}}e^{-\overline{\mathcal{K}}(\mathcal{T} + 2\mathfrak{T})}\alpha_l. \quad (4.12)$$

Note that there is now at least one additional agent $I \in \mathcal{N} \setminus \mathcal{I}_l$ with $y_{I[k]}(t_0 + l(\mathcal{T} + \mathfrak{T})) \leq \overline{\Upsilon_k}(t_0) - \alpha_{l+1}$.

The iteration step ends with updating the parameters. In the first case, i.e. $y_{i_{R_l}[k]}(t_0) > \overline{\Upsilon_k}(t_0) - \alpha_l$, we have $\alpha_{l+1} = \beta_l = e^{-\overline{\mathcal{K}}(\mathcal{T} + \mathfrak{T})}\alpha_l$, which is always larger than (4.12). Therefore, we take (4.12) for all $2N$ iterations. For the set $\mathcal{I}_l$, we have $\mathcal{I}_{l+1} = \mathcal{I}_l$ in the first case and $\mathcal{I}_{l+1} = \mathcal{I}_l + \{I\}$ in the second case, where $I \in \mathcal{N} \setminus \mathcal{I}_l$. Since the second case appears at least N times, we have $\mathcal{I}_{2N+1} = \mathcal{N}$. Hence, all agents $i \in \mathcal{N}$ satisfy $y_{i[k]}(t_0 + 2N(\mathcal{T} + \mathfrak{T})) \leq \overline{\Upsilon_k}(t_0) - \alpha_{2N+1}$, where

$$\alpha_{2N+1} = \prod_{l=1}^{2N}\left((1 - e^{-(\underline{\mathcal{K}}_l + \overline{\mathcal{K}})h_{DW}})\frac{\underline{\mathcal{K}}_l}{\underline{\mathcal{K}}_l + \overline{\mathcal{K}}}e^{-\overline{\mathcal{K}}(\mathcal{T} + 2\mathfrak{T})}\right)\frac{1}{2}\widehat{\Upsilon_k}(t_0).$$

Then, we have to wait for an additional interval of length $\mathcal{T}$ such that $y_{i[k]}(t_0 + T_{\text{iter}} + \eta) \leq \overline{\Upsilon_k}(t_0) - \varepsilon(\widehat{\Upsilon_k}(t_0))$ for all $\eta \in [-\mathcal{T}, 0]$ where $T_{\text{iter}} = 2N(\mathcal{T} + \mathfrak{T}) + \mathcal{T}$ and

$$\varepsilon(\widehat{\Upsilon_k}(t_0)) = \prod_{l=1}^{2N}\left((1 - e^{-(\underline{\mathcal{K}}_l + \overline{\mathcal{K}})h_{DW}})\frac{\underline{\mathcal{K}}_l}{\underline{\mathcal{K}}_l + \overline{\mathcal{K}}}e^{-\overline{\mathcal{K}}(\mathcal{T} + 2\mathfrak{T})}\right)e^{-\overline{\mathcal{K}}\mathcal{T}}\frac{1}{2}\widehat{\Upsilon_k}(t_0) > 0,$$

i.e. we have $\widehat{\Upsilon_k}(t_0 + T_{\text{iter}}) - \widehat{\Upsilon_k}(t_0) \leq -\varepsilon(\widehat{\Upsilon_k}(t_0))$, see (4.10). Note that ε is a nonlinear function of $\widehat{\Upsilon_k}(t_0)$ because $\underline{\mathcal{K}}_l$ also depends on $\widehat{\Upsilon_k}(t_0)$ as it has to satisfy $\underline{\mathcal{K}}_l \leq \underline{\mathcal{K}}((1 - \frac{\underline{\mathcal{K}}_l}{\underline{\mathcal{K}}_l + \overline{\mathcal{K}}})\beta_l)$. In addition, this $\underline{\mathcal{K}}_l$ also satisfies $\underline{\mathcal{K}}_l \leq \underline{\mathcal{K}}((1 - \frac{\underline{\mathcal{K}}_l}{\underline{\mathcal{K}}_l + \overline{\mathcal{K}}})\beta^*_l)$ for any $\beta^*_l \geq \beta_l$ because $\underline{\mathcal{K}}$ is non-decreasing. Therefore, $\varepsilon(\cdot)$ is also non-decreasing because larger $\widehat{\Upsilon_k}(t_0)$ lead to larger (or at least not smaller) β_l and this allows at least for the same $\underline{\mathcal{K}}_l$.

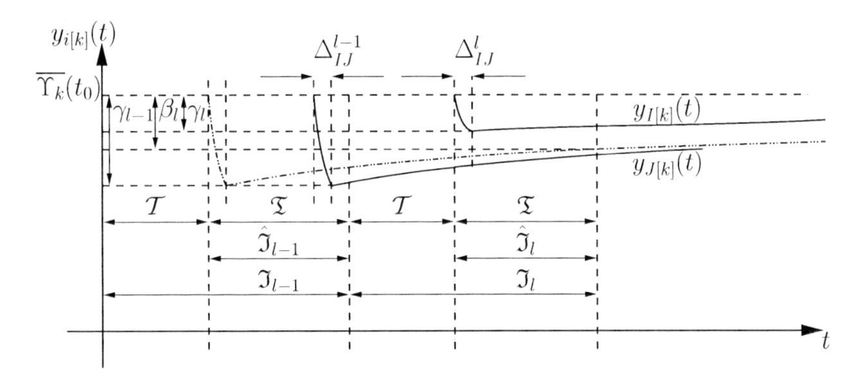

(a) Alternating intervals $\mathcal{T}$ and $\mathfrak{T}$.

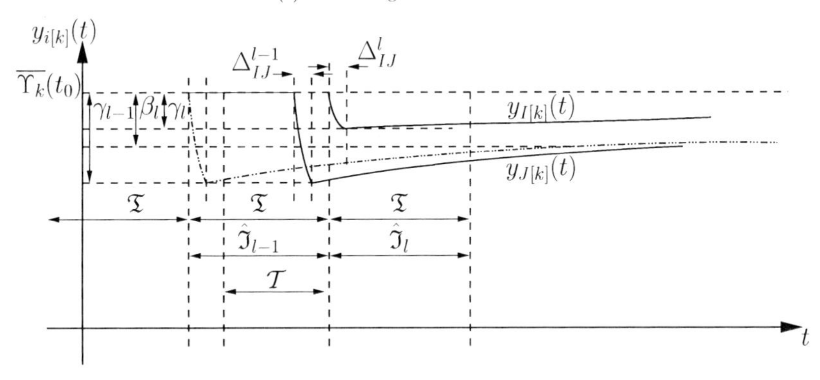

(b) Consecutive intervals $\mathfrak{T}$.

Figure 4.3: Schematic illustrations for alternating and consecutive intervals as discussed in the proof of Theorem 4.5.

At this point, we explain briefly the structure of the intervals $\mathfrak{I}_l$ with alternating intervals of length $\mathcal{T}$ and $\mathfrak{T}$, see Figure 4.3(a) for an illustration. In order to determine γ_l, we consider an agent I that is influenced by another agent J during some interval $\Delta^l_{IJ} \subset \hat{\mathfrak{I}}_l$. Our previous arguments require $\mathbb{T}_{IJ}(y_{J,t[k]}) \leq \overline{\Upsilon_k}(t_0)-\beta_l$ for all $t \in \Delta^l_{IJ}$. If we consider the case where Δ^l_{IJ} is at the beginning of the interval $\hat{\mathfrak{I}}_l$, see Figure 4.3(a), we see that there must be an interval of length $\mathcal{T}$ before the interval $\hat{\mathfrak{I}}_l$ where $y_{J[k]} \leq \overline{\Upsilon_k}(t_0)-\beta_l$. Otherwise, $\mathbb{T}_{IJ}(y_{J,t[k]}) \leq \overline{\Upsilon_k}(t_0) - \beta_l$ is not guaranteed. Yet, $y_{J[k]} \leq \overline{\Upsilon_k}(t_0) - \beta_l$ might not be true if we concatenate intervals of length $\mathfrak{T}$ as in the undelayed case, see Lin (2006) and Figure 4.3(b), because the corresponding parent J might be the agent that was added to the set $\mathcal{I}_{l-1}$ in the last interval, i.e. it was attracted by a parent in $\mathcal{I}_{l-1}$ during the interval Δ^{l-1}_{IJ}. This interval Δ^{l-1}_{IJ} can be anywhere inside $\hat{\mathfrak{I}}_{l-1}$. The dash-dotted line in Figure 4.3(b) shows the upper bound of the solution of $y_{J[k]}$ if Δ^{l-1}_{IJ} is at the beginning of $\hat{\mathfrak{I}}_{l-1}$, see also Figure 4.2. The solid line is the upper bound of the solution of $y_{J[k]}$ if Δ^{l-1}_{IJ} is toward the end of $\hat{\mathfrak{I}}_{l-1}$. We see immediately that the solid line does not satisfy $y_{J[k]} \leq \overline{\Upsilon_k}(t_0) - \beta_l$ for all t in an interval of length $\mathcal{T}$ before the interval $\hat{\mathfrak{I}}_l$. Therefore, we always have to alternate intervals of length $\mathcal{T}$ and of length $\mathfrak{T}$.

Conclusion For any $\upsilon \in \mathbb{N}$ and for all t_0, we have

$$\begin{aligned}
\widehat{\Upsilon_k}(t_0 + \upsilon T_{\text{iter}}) &\leq \widehat{\Upsilon_k}(t_0) - \varepsilon(\widehat{\Upsilon_k}(t_0)) \\
&\quad - \varepsilon(\widehat{\Upsilon_k}(t_0 + T_{\text{iter}})) - \ldots - \varepsilon(\widehat{\Upsilon_k}(t_0 + (\upsilon-1)T_{\text{iter}})) \\
&\leq \widehat{\Upsilon_k}(t_0) - \upsilon\varepsilon(\widehat{\Upsilon_k}(t_0 + (\upsilon-1)T_{\text{iter}})) \\
&\leq \widehat{\Upsilon_k}(t_0) - \upsilon\varepsilon(\widehat{\Upsilon_k}(t_0 + \upsilon T_{\text{iter}})),
\end{aligned}$$

because $\widehat{\Upsilon_k}$ is non-increasing and ε is non-decreasing. We conclude that, for given $\widehat{\Upsilon_k}(t_0)$ and any arbitrarily small $\delta > 0$, there exists a $T^* = \upsilon T_{\text{iter}}$ with υ sufficiently large such that $\widehat{\Upsilon_k}(t_0 + t) < \delta$ for all $t \geq T^*$. This can be shown easily by contradiction. Assume $\widehat{\Upsilon_k}(t_0 + \upsilon T_{\text{iter}}) \geq \delta$ for all υ. Then, the above inequalities provide $\delta \leq \widehat{\Upsilon_k}(t_0) - \upsilon\varepsilon(\delta)$ which is false for sufficiently large υ because $\varepsilon(\delta) > 0$. Therefore, $\lim_{t\to\infty} \widehat{\Upsilon_k}(t) = 0$ for any k. Hence, rendezvous is achieved. □

Theorem 4.5 extends the main results of Lin (2006) from nonlinear MAS without delays to nonlinear RD1 MAS with heterogeneous delays taking all three delay models into account. It is very interesting to see that rendezvous is achieved for arbitrary large but bounded delays under exactly the same assumptions as in the undelayed case. In fact, Lin (2006) provides several counter-examples illustrating that rendezvous is not guaranteed if these assumptions are relaxed. Clearly, these counter examples also apply for the MAS with delays discussed here because the delay robustness analysis also includes the zero delay case.

Theorem 4.5 holds of course also for constant topologies. This is expressed in the following corollary, that is stated here for completeness without proof.

Corollary 4.6 (Rendezvous in nonlinear RD1 MAS with constant topology). *A MAS with agent dynamics* (4.1) *that satisfy Assumption 4.1 and with interconnection* (4.3) *where $\tilde{k}_{ij}$ satisfy Design Condition 4.3 achieves rendezvous asymptotically on networks of arbitrary size $N \in \mathbb{N}$, with arbitrary quasi-strongly connected directed topologies, and arbitrary bounded heterogeneous constant, time-varying, or distributed delays.*

4.3 Particular Multi-Agent Systems with Relative Degree One

Design Condition 4.3 and in particular Condition (4.5) implicitly impose additional assumptions on the agent dynamics. In this section, we consider special cases of MAS (4.1) and reformulate the conditions in Theorem 4.5 for these cases. In particular, we consider nonlinear input affine RD1 MAS, Euler-Lagrange MAS, and linear RD1 MAS.

4.3.1 Nonlinear, Input Affine Relative Degree One Multi-Agent Systems

Consider nonlinear, input affine RD1 MAS consisting of N agents with dynamics

$$\Sigma_i : \begin{cases} \dot{x}_i(t) &= f_i(x_i(t)) + g_i(x_i(t))u_i(t) \\ y_i(t) &= h_i(x_i(t)) \end{cases}, \qquad i \in \mathcal{N}, \tag{4.13}$$

where $x_i(t) \in \mathbb{R}^{n_i}, u_i(t) \in \mathbb{R}^m, y_i(t) \in \mathbb{R}^m$ are the state, input, and output of agent i, respectively. The initial condition of (4.13) with interconnection (4.3) is $\varphi \in \mathcal{C}_{\sum n_i} = \mathcal{C}([-\mathcal{T}, 0], \mathbb{R}^{\sum n_i})$. The functions $f_i, g_i, h_i, \tilde{k}_{ij}$, and k_{ij} are sufficiently smooth such that the closed loop system has piecewise continuously differentiable solutions, even for piecewise continuous time-varying delays.

Assumption 4.1 transforms into the following:

Assumption 4.7. *The agents* (4.13) *have stable zero dynamics and uniform relative degree one, i.e. $\frac{\partial h_i}{\partial x_i} g_i(x_i)$ is non-singular for all $x_i \in \mathbb{R}^{n_i}$.*

Moreover, we repeatedly use the following Design Condition, which replaces Design Condition 4.3

Design Condition 4.8. *The continuous coupling functions $k_{ij} : \mathbb{R}^m \to \mathbb{R}^m$ are such that $k_{ij}(z) = \overline{k_{ij}}(\|z\|)\frac{z}{\|z\|}$ with nonlinear, continuous gain $\overline{k_{ij}} : \mathbb{R}_0^+ \to \mathbb{R}_0^+$ that satisfies $k_{ij}(0) = \overline{k_{ij}}(0) = 0$ and $\overline{k_{ij}}(\|z\|) > 0$ for all $\|z\| \neq 0$, i.e. $\overline{k_{ij}}$ is strictly passive.*

Then, we have the following result.

Corollary 4.9 (Rendezvous in nonlinear, input affine, RD1 MAS with input-output linearization). *A MAS with agent dynamics* (4.13) *that satisfy Assumption 4.7 and with interconnection* (4.3) *achieves rendezvous asymptotically on networks of arbitrary size $N \in \mathbb{N}$, with arbitrary uniformly quasi-strongly connected directed switching topologies, and arbitrary bounded heterogeneous constant, time-varying, or distributed delays if the* *coupling functions $\tilde{k}_{ij}$ satisfy*

$$\tilde{k}_{ij}(x_i(t), y_i(t) - \mathbb{T}_{ij}(y_{j,t})) = \left(\frac{\partial h_i}{\partial x_i} g_i(x_i)\right)^{-1} \left(\frac{\partial h_i}{\partial x_i} f_i(x_i) + k_{ij}(y_i(t) - \mathbb{T}_{ij}(y_{j,t}))\right), \tag{4.14}$$

where k_{ij} satisfies Design Condition 4.8.

Proof. We show that (4.14) guarantees that (4.5) is satisfied. Note first that the inverse of $\frac{\partial h_i}{\partial x_i} g_i(x_i)$ exists because the agents have relative degree one, see Assumption 4.7. Using (4.14) in (4.3) results in

$$\begin{aligned} u_i(t) &= -\sum_{j=1}^{N} \frac{a_{ij}(t)}{d_i(t)} \left(\frac{\partial h_i}{\partial x_i} g_i(x_i)\right)^{-1} \left(\frac{\partial h_i}{\partial x_i} f_i(x_i) + k_{ij}(y_i(t) - \mathbb{T}_{ij}(y_{j,t}))\right) \\ &= -\left(\frac{\partial h_i}{\partial x_i} g_i(x_i)\right)^{-1} \left(\frac{\partial h_i}{\partial x_i} f_i(x_i) + \sum_{j=1}^{N} \frac{a_{ij}(t)}{d_i(t)} k_{ij}(y_i(t) - \mathbb{T}_{ij}(y_{j,t}))\right), \end{aligned}$$

because $d_i(t) = \sum_{j=1}^{N} a_{ij}(t)$. Thus, we have

$$\begin{aligned} \dot{y}_i(t) = \frac{\partial h_i}{\partial x_i} f_i(x_i) + \frac{\partial h_i}{\partial x_i} g_i(x_i) u_i(t) &= -\sum_{j=1}^{N} \frac{a_{ij}(t)}{d_i(t)} k_{ij}(y_i(t) - \mathbb{T}_{ij}(y_{j,t})) \\ &= -\sum_{j=1}^{N} \frac{a_{ij}(t)}{d_i(t)} \overline{k_{ij}}(\|y_i(t) - \mathbb{T}_{ij}(y_{j,t})\|) \frac{y_i(t) - \mathbb{T}_{ij}(y_{j,t})}{\|y_i(t) - \mathbb{T}_{ij}(y_{j,t})\|}, \end{aligned} \tag{4.15}$$

with Design Condition 4.8. Since $\overline{k_{ij}}$ is continuous and strictly passive, we have

$$\dot{y}_i(t) = \sum_{j \in \mathcal{N}_i} \widehat{k_{ij}}(t) \left(\mathbb{T}_{ij}(y_{j,t}) - y_i(t)\right), \quad \text{for all } t \geq 0,$$

where $\widehat{k_{ij}}(t) = \frac{a_{ij}(t)}{d_i(t)} \frac{\overline{k_{ij}}(\|y_i(t) - \mathbb{T}_{ij}(y_{j,t})\|)}{\|y_i(t) - \mathbb{T}_{ij}(y_{j,t})\|} \geq 0$. Therefore, Lemma 4.4 holds, i.e. the set $\Upsilon(t)$ is non-increasing. Moreover, we have

$$\sum_{j \in \mathcal{N}_i(t)} \widehat{k_{ij}}(t) \leq \overline{\mathcal{K}} = \max_{i,j} \sup_{t \geq 0} \frac{\overline{k_{ij}}(\|y_i(t) - \mathbb{T}_{ij}(y_{j,t})\|)}{\|y_i(t) - \mathbb{T}_{ij}(y_{j,t})\|},$$

for all $t \geq 0$. This supremum is finite because k_{ij} and $\overline{k_{ij}}$ are continuous, $k_{ij}(0) = 0$, and $\|y_i(t) - \mathbb{T}_{ij}(y_{j,t})\|$ is bounded because $\Upsilon(0)$ is positively invariant. In addition, we have

$$\widehat{k_{ij}}(t) \geq \underline{\mathcal{K}}(\eta) = \min_{i,j} \inf_{\|z\| \geq \eta} \frac{a_{ij}(t)}{d_i(t)} \frac{\overline{k_{ij}}(\|z\|)}{\|z\|},$$

for all $i \in \mathcal{N}, j \in \mathcal{N}_i(t)$, where $\underline{\mathcal{K}}$ is non-decreasing by construction and $\underline{\mathcal{K}}(\eta) > 0$ for $\eta > 0$ because (i) $a_{ij}(t)$ and $d_i(t)$ are constant as long as the graph does not switch and there is only a finite number of graphs $\mathcal{G}_p, p \in \mathcal{P}$, (ii) $\overline{k_{ij}}(\|z\|) > 0$ for all $\|z\| \neq 0$, and (iii) $\|z\|$ is again upper bounded by $\Upsilon(0)$. □

The coupling functions (4.14) use an input-output linearization, see Khalil (2002) for example. Therefore, Corollary 4.9 can only be applied if the dynamics f_i, g_i, and h_i are known. If this is not the case, we might ask for additional assumptions on the agent dynamics (4.13) such that an input-output linearization is not required. Such conditions are summarized in the following assumption:

Assumption 4.10. *The agents* (4.13) *have an integrating behavior from input to output, i.e.* $\frac{\partial h_i}{\partial x_i} f_i(x_i) = 0$ *for all* x_i*, and every element of the input influences only one element of the output, i.e.* $\frac{\partial h_i}{\partial x_i} g_i(x_i)$ *is a diagonal matrix (possibly after an appropriate renumbering of the inputs and outputs). Moreover, we assume that there exist* $\underline{\epsilon} > 0, \overline{\epsilon} > 0$ *such that* $\underline{\epsilon} I \leq \frac{\partial h_i}{\partial x_i} g_i(x_i(t)) \leq \overline{\epsilon} I$ *for all* $t \geq 0$.

The last assumption is necessary to derive time-invariant bounds $\overline{\mathcal{K}}, \underline{\mathcal{K}}$. With this assumption, we can state the following result:

Corollary 4.11 (Rendezvous in nonlinear, input affine, RD1 MAS with integrating behavior)**.** *A MAS with agent dynamics* (4.13) *that satisfy Assumptions 4.7 and 4.10 and with interconnection* (4.4) *where* k_{ij} *satisfy Design Condition 4.8 achieves rendezvous asymptotically on networks of arbitrary size* $N \in \mathbb{N}$*, with arbitrary uniformly quasi-strongly connected directed switching topologies, and arbitrary bounded heterogeneous constant, time-varying, or distributed delays.*

Proof. Due to Assumption 4.10, we have

$$\dot{y}_i(t) = -\frac{\partial h_i}{\partial x_i} g_i(x_i) \sum_{j=1}^{N} \frac{a_{ij}}{d_i} \overline{k_{ij}}(\|y_i(t) - \mathbb{T}_{ij}(y_{j,t})\|) \frac{y_i(t) - \mathbb{T}_{ij}(y_{j,t})}{\|y_i(t) - \mathbb{T}_{ij}(y_{j,t})\|}, \tag{4.16}$$

and the matrix $\frac{\partial h_i}{\partial x_i} g_i(x_i)$ is diagonal and has strictly positive and bounded entries on the diagonal for all $t \geq 0$. This implies that (4.5) holds using similar arguments as in the proof of Corollary 4.9. □

Assumption 4.10 is in fact quite restrictive for general dynamics with relative degree one. We explain briefly why this assumption is reasonable, e.g., for a set of robots. First of all, (4.5) requires that the robots are holonomic, i.e. they are able to move in any direction using omnidirectional drives. This is a standard assumption for MAS and there are only few results for non-holonomic MAS, e.g. Dong and Farrell (2008); Sepulchre et al. (2008). If the robots have omnidirectional drives, then there is usually a one-to-one mapping from the inputs to outputs, e.g. from desired vertical speed and horizontal speed to an according movement in vertical and horizontal direction. Hence, the conditions on $\frac{\partial h_i}{\partial x_i} g_i(x_i)$ are reasonable in this situation. The integrating behavior requires that the internal states of the robot only affect the gain from input to output. This integrating behavior is indeed quite reasonable for consensus and rendezvous, compare also our previous arguments in Section 4.1.2.

Clearly, Assumption 4.10 is also satisfied if the input-output dynamics are just an integrator, i.e. $\dot{y}_i(t) = u_i(t)$. This is discussed in the Section 4.3.3.

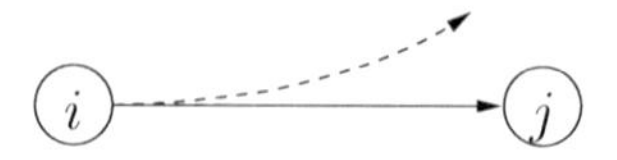

Figure 4.4: Motion of robot i heading for robot j with compensation of centrifugal forces (solid line) and without compensation (dashed line).

4.3.2 Euler-Lagrange Systems

In this subsection, we show how Theorem 4.5 can be applied to Euler-Lagrange systems. Euler-Lagrange systems describe a broad class of nonlinear electro-mechanical systems like robots, see for example Ortega et al. (1998). We consider fully actuated Euler-Lagrange systems, where we assume that the gravitational forces are compensated by some local internal controller. Agent i's dynamics are thus given by

$$M_i(y_i)\ddot{y}_i + C_i(y_i, \dot{y}_i)\dot{y}_i = u_i, \qquad i \in \mathcal{N}, \tag{4.17}$$

where $y_i \in \mathbb{R}^m$ is the vector of generalized configuration coordinates of agent i and $u_i \in \mathbb{R}^m$ is the vector of generalized forces acting on system i. The cooperative control task is to achieve rendezvous in the generalized velocities, i.e. $\lim_{t\to\infty} \dot{y}_i(t) - \dot{y}_j(t) = 0$. Note that the agents have relative degree one from u_i to $\dot{y}_i$. The inertia matrices $M_i(y_i) \in \mathbb{R}^{m\times m}$ is positive definite and the vector of centrifugal and Coriolis forces $C_i(y_i, \dot{y}_i)\dot{y}_i \in \mathbb{R}^m$ is such that $\frac{d}{dt}M_i(y_i) - 2C_i$ is skew symmetric.

Following Corollary 4.9, a feedback (4.3) can be designed that achieves rendezvous based on an input-output linearization, i.e.

$$\tilde{k}_{ij}(\dot{y}_i(t), y_i(t), \dot{y}_i(t) - \mathbb{T}_{ij}(\dot{y}_{j,t})) = -C_i(y_i, \dot{y}_i)\dot{y}_i + M_i(y_i)k_{ij}(\dot{y}_i(t) - \mathbb{T}_{ij}(\dot{y}_{j,t})),$$

where k_{ij} satisfies Design Condition 4.8. This coupling function guarantees rendezvous for directed, uniformly quasi-strongly connected, switching graphs.

In fact, Euler-Lagrange MAS with delays achieves rendezvous also with interconnection (4.4), where k_{ij} satisfies Design Condition 4.8, if the graph is constant, balanced, and strongly connected. This can be shown following the proof in Chopra et al. (2008) which uses sums of Lyapunov-Krasovskii functionals similar to the proof of Theorem 5.3 in Chapter 5. In this case, the feedback avoids an input-output linearization but requires stronger assumptions on the graph topology.

Motivated by the results in Chopra et al. (2008), one might ask for a similar feedback that avoids input-output linearization and achieves rendezvous on directed, uniformly quasi-strongly connected, switching graphs. This is however anything but straightforward, as explained in the following. Consider a robot with three degrees of freedom in the plain, i.e. vertical and horizontal movement as well as rotation or heading. If this robot is rotating and the input is heading for its parent robot, then the centrifugal forces will drive the robot on a curve instead of a straight line, see Figure 4.4 for an illustration. The robot will move on a straight line toward its parent robot only if the centrifugal forces are appropriately compensated, e.g. by an input-output linearization. This simple example illustrates why general Euler-Lagrange MAS without input-output linearization usually do not satisfy Design Condition 4.3.

4.3.3 Linear, Relative Degree One Multi-Agent Systems

Now, we consider general linear RD1 MAS of the following form:

$$\Sigma_i : \begin{cases} \dot{x}_i(t) &= A_i x_i(t) + B_i u_i(t) \\ y_i(t) &= C_i x_i(t) \end{cases}, \qquad i \in \mathcal{N}, \tag{4.18}$$

where $x_i(t) \in \mathbb{R}^{n_i}, u_i(t) \in \mathbb{R}^m, y_i(t) \in \mathbb{R}^m$ are the state, input, and output of agent i, respectively, and $A_i : \mathbb{R}^{n_i \times n_i}, B_i : \mathbb{R}^{n_i \times m}$ and $C_i : \mathbb{R}^{m \times n_i}$. The initial condition is $\varphi \in \mathcal{C}_{\sum n_i} = \mathcal{C}([-\mathcal{T}, 0], \mathbb{R}^{\sum n_i})$. Obviously, (4.18) is the linear counterpart of (4.13). Assumption 4.7 transforms into $C_i B_i$ being nonsingular for all i. Thus, we have the following result which follows directly from Corollary 4.9 and is therefore stated without proof:

Corollary 4.12 (Rendezvous in linear RD1 MAS). *A MAS with agent dynamics* (4.18)*, where $C_i B_i$ is nonsingular and the zero dynamics are stable, achieves rendezvous asymptotically with interconnection* (4.3) *on networks of arbitrary size $N \in \mathbb{N}$, with arbitrary uniformly quasi-strongly connected directed switching topologies, and arbitrary bounded heterogeneous constant, time-varying, or distributed delays if the coupling functions $\tilde{k}_{ij}$ satisfy*

$$\tilde{k}_{ij}(x_i(t), y_i(t) - \mathbb{T}_{ij}(y_{j,t})) = (C_i B_i)^{-1} \left(C_i A_i x_i(t) + k_{ij}(y_i(t) - \mathbb{T}_{ij}(y_{j,t})) \right), \tag{4.19}$$

where k_{ij} satisfy Design Condition 4.8.

The standard model for rendezvous is obtained for $C_i A_i = 0$ and $C_i B_i = 1$:

$$\dot{y}_i(t) = u_i(t), \tag{4.20}$$

the single integrator MAS, e.g. Olfati-Saber et al. (2007); Ren and Beard (2008). Finally, the following corollary presents conditions for this MAS with feedback (4.4) to achieve rendezvous independent of delay. This is an extension of Corollary 3.9 toward nonlinear coupling functions, directed, switching graphs and time-varying or distributed delays.

Corollary 4.13 (Rendezvous in single integrator MAS). *A single integrator MAS* (4.20) *with interconnection* (4.4) *where k_{ij} satisfy Design Condition 4.8 achieves rendezvous asymptotically on networks of arbitrary size $N \in \mathbb{N}$, with arbitrary uniformly quasi-strongly connected directed switching topologies, and arbitrary bounded heterogeneous constant, time-varying, or distributed delays.*

The proof follows directly from Corollary 4.9 and is therefore omitted.

In this section, we presented conditions on the coupling functions $\tilde{k}_{ij}$ and k_{ij} such that special cases of MAS (4.1) reach a rendezvous independent of delay on uniformly quasi-strongly connected directed switching graphs. The next section is dedicated to a practical application: the synchronization of Kuramoto oscillators.

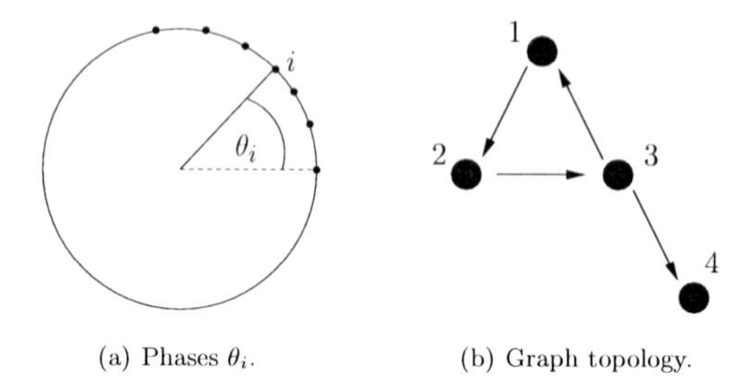

(a) Phases θ_i.

(b) Graph topology.

Figure 4.5: Phases θ_i of different Kuramoto oscillators on a circle, see (4.21), and graph topology of the network considered in Section 4.4.2.

4.4 Synchronization of Kuramoto Oscillators

We complete this chapter illustrating how the developed methods can be used to prove synchronization of Kuramoto oscillators. The Kuramoto oscillator is particularly interesting because it is not straightforward to see how the methods developed in this chapter apply for this MAS.

The *Kuramoto oscillator* was originally proposed in Kuramoto (1984). In physics, it is one of the standard models to study synchronization of oscillators, see Strogatz (2000) for an overview. A fundamental assumption of this model requires that all oscillators expose identical, asymptotically attracting limit cycles. Moreover, the coupling between the agents may not destabilize these limit cycles. If these assumptions are satisfied, then the synchronization can be analysed using a phase-model on the limit cycle. For simplicity, it is assumed that the limit cycle is the unit circle, see Figure 4.5(a). Each agent oscillates with identical eigenfrequency $\omega_e \in \mathbb{R}$ on the unit circle. The phase $\theta_i(t) \in \mathbb{R}$ of oscillator i is driven by the following dynamics

$$\dot{\theta}_i(t) = \omega_e - K \sum_{j=1}^{N} \frac{a_{ij}}{d_i} \sin(\theta_i(t) - \theta_j(t)), \tag{4.21}$$

where $K > 0$ is the coupling gain, a_{ij} are the elements of the adjacency matrix of the underlying graph, and d_i is the in-degree of node i. For simplicity, we assume a constant graph in this section.

Networks of Kuramoto oscillators are often used to model synchronization of oscillators in different fields of biology, physics, and engineering. As a practical application, we may think of a network of agents, e.g. sensors, that try to synchronize their local clock pulses. Clock synchronization in packet-switched communication network (PSCN) is in fact a very critical and difficult problem in industrial networked control systems, see for example Lupas Scheiterer et al. (2009). The PSCN between the agents is modeled in the simulations using MATLAB® Simulink® and the SimEvents® toolbox (MathWorks, 2006). For the analysis, we assume that the packet-delay and packet-loss can be modeled in a distributed delay framework, see Münz et al. (2009b) for details. Therefore, the oscillator model (4.21) is extended to

$$\dot{\theta}_i(t) = \omega_e - K \sum_{j=1}^{N} \frac{a_{ij}}{d_i} \sin\left(\theta_i(t) - \int_0^{\mathcal{T}} \phi(\eta)\theta_j(t-\eta)d\eta\right), \tag{4.22}$$

where ϕ describe the delay distribution of the communication channels. For simplicity, we assume all channels to have the same statistics, i.e. $\phi, \mathcal{T}$ instead of $\phi_{ij}, \mathcal{T}_{ij}$. A similar model with constant delays instead of distributed delays has been studied in Yeung and Strogatz (1999); Earl and Strogatz (2003).

The aim of our analysis is to provide sufficient conditions for synchronization, i.e. such that $\lim_{t\to\infty}(\theta_i(t) - \theta_j(t)) = 0$ for all $i, j \in \mathcal{N}$. We restrict the presentation to identical oscillators, i.e. all oscillators have the same eigenfrequency ω_e. The more general case with non-identical oscillators has been studied in Papachristodoulou et al. (2010) and our previous publications Schmidt et al. (2009); Papachristodoulou et al. (2010). We first analyse (4.22) using the results of this chapter in Subsection 4.4.1. These findings are then compared to simulations of Kuramoto oscillators exchanging phase information over a PSCN modeled in Simulink in Subsection 4.4.2.

4.4.1 Analysis of Kuramoto Oscillators

The Kuramoto oscillator with nonzero frequency ω_e cannot be separated into agent dynamics (4.1) and interconnection (4.3) satisfying the corresponding assumptions and design conditions above. We have to transform the Kuramoto dynamics in a rotating coordinate system. To this end, we identify a consistency constraint as proposed in Yeung and Strogatz (1999); Papachristodoulou et al. (2010). If the oscillators synchronize, then there exist $\overline{\omega}, \overline{\theta} \in \mathbb{R}$ such that $\lim_{t\to\infty}(\theta_i(t) - \overline{\omega}t) = \overline{\theta}$, for all i. Thereby, $\overline{\omega}$ is the frequency of the synchronized oscillators. Such a synchronized solution $\theta_i(t) = \overline{\omega}t + \overline{\theta}$ of (4.22) exists only if the following consistency constraint holds

$$\overline{\omega} = \omega_e - K \sin(\overline{\omega}\,\overline{\tau}), \tag{4.23}$$

where $\overline{\tau} = \int_0^{\mathcal{T}} \eta\phi(\eta)d\eta$ and we use $\int_0^{\mathcal{T}} \phi(\eta)d\eta = 1$. The interested reader is referred to Yeung and Strogatz (1999) for a detailed study of the solutions of (4.23), in particular with respect to the stability of synchronized and incoherent states for specific values of $\overline{\tau}$ and K.

If the consistency constraint (4.23) holds, we can transform the Kuramoto dynamics in a rotating coordinate system with rotation frequency $\overline{\omega}$. The phase y_i in the new coordinate system is $y_i(t) = \theta_i(t) - \overline{\omega}t$ for all i and we obtain

$$\begin{aligned}\dot{y}_i(t) &= \omega_e - K \sum_{j=1}^{N} \frac{a_{ij}}{d_i} \sin\left(\theta_i(t) - \int_0^{\mathcal{T}} \phi(\eta)\theta_j(t-\eta)d\eta\right) - \overline{\omega}\\ &= -K \sum_{j=1}^{N} \frac{a_{ij}}{d_i} \left(\sin\left(y_i(t) - \int_0^{\mathcal{T}} \phi(\eta)y_j(t-\eta)d\eta + \overline{\omega}\,\overline{\tau}\right) - \sin\left(\overline{\omega}\,\overline{\tau}\right)\right). \end{aligned} \tag{4.24}$$

We will show in the following that y_i achieves a consensus, i.e. the oscillators synchronize, for appropriate initial conditions, coupling gain K, and delay distribution ϕ. Note that

$$\bar{k}(z) = \sin(z + \bar{\omega}\mathcal{T}) - \sin(\bar{\omega}\mathcal{T}),$$

is locally passive if $\left.\frac{d\bar{k}}{dz}\right|_{z=0} = \cos(\bar{\omega}\mathcal{T}) > 0$. Moreover, we have $\bar{k}(z) = 0$ if $z = 2k\pi$ or $z = (2k+1)\pi - 2\bar{\omega}\mathcal{T}$ for all $k \in \mathbb{Z}$. With this result, we specify the domain $[-\bar{z}, \bar{z}]$ such that $z\bar{k}(z) > 0$ for all $z \in [-\bar{z}, \bar{z}]$ as follows

$$\bar{z} < \min_{k\in\mathbb{Z}} |(2k+1)\pi - 2\bar{\omega}\mathcal{T}|, \tag{4.25}$$

see also Papachristodoulou et al. (2010) and remember that $\cos(\bar{\omega}\mathcal{T}) > 0$. Summarizing, the Kuramoto oscillators (4.22) synchronize if the self-consistency condition (4.23) holds with parameter $\bar{\omega}$ such that $\cos(\bar{\omega}\mathcal{T}) > 0$ and if the initial condition satisfies $|y_i(\eta)| = |\theta_i(\eta) - \bar{\omega}\eta - \tilde{\theta}| \leq \frac{\bar{z}}{2}$ for all $\eta \in [-\mathcal{T}, 0]$ and some $\tilde{\theta} \in \mathbb{R}$. This follows directly from Corollary 4.13.

4.4.2 Simulation Results

The simulations are performed using MATLAB® Simulink®. The PSCN is simulated with the SimEvents® toolbox (MathWorks, 2006). The continuous-time signals transmitted over the PSCN are first sampled with sampling period T_s. Then, every ten consecutive sampled values are stored in one packet, and a packet is sent over the network every $T_p = 10T_s$ seconds. We compare two networks with different bandwidth. The first one is slower and allows for transmissions every $T_{p1} = 0.05s$, i.e. the sampling time is $T_{s1} = 0.005s$. The second network is faster with double packet rate and half the sampling time, i.e. $T_{p2} = 0.025s$ and $T_{s2} = 0.0025s$. The stochastic packet delays are simulated using a queue and a server. The required server processing time for each packet is a gamma-distributed random process with parameters α and β. The shape parameter is set to $\alpha = 5$ and the scaling parameter is $\beta_1 = 0.005$ for the slow network with $T_{s1} = 0.005$ and $\beta_2 = 0.0025$ for the fast network with $T_{s2} = 0.0025$. Note that the sampling times and scaling parameters are strongly related because the server processes the packets faster or slower depending on β. Note however that $\beta_i = T_{si}$ is actually not necessary. The server processing time distribution and its parameters are chosen such that a realistic packet delay distribution is achieved, as shown exemplarily in the histograms in Figure 4.6. Corresponding measurements from real communication networks are presented for example in Moyne and Tilbury (2007); Lopez et al. (2006); Roesch et al. (2005); Salza et al. (2000). At the receiver side, a continuous-time piecewise constant signal is reconstructed by extracting the ten sampled values one after the other from the packets. If the difference of the arrival times of two consecutive packets is more than T_P, the last value from the old packet is held until the new packet arrives. If this difference is less than T_P, then the remaining data from the old packet is dismissed and the data from the new packet is read out immediately. The average delay of the data transmitted over the network consists of two parts: (i) a constant delay T_p induced by

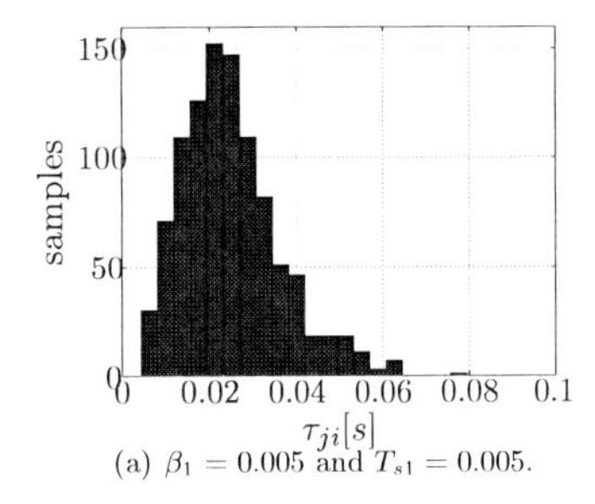

(a) $\beta_1 = 0.005$ and $T_{s1} = 0.005$.

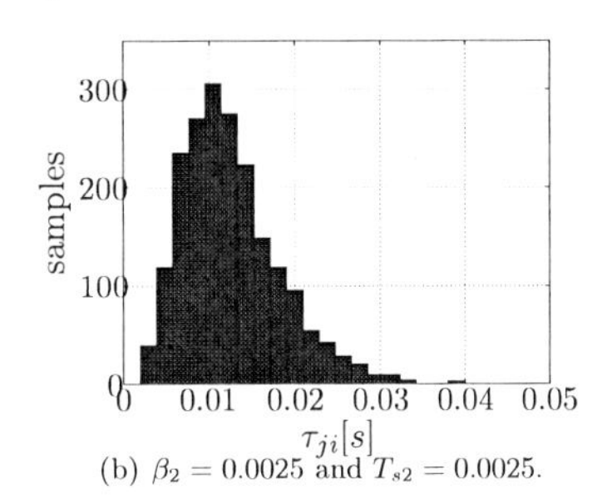

(b) $\beta_2 = 0.0025$ and $T_{s2} = 0.0025$.

Figure 4.6: Exemplary histograms for simulated packet delays depending on the scaling parameter β and the sampling period T_s.

merging 10 samples in one packet and (ii) a stochastic delay from the queue and the server, approximately the average of the packet delays shown in Figure 4.6. For the slow network, this sums up to an average delay $\overline{\tau}_1 \approx 0.075s$ and for the fast network to $\overline{\tau}_2 \approx 0.0375s$. Similarly, the maximal delays are $\mathcal{T}_1 \approx 0.13s$ for the slow PSCN and $\mathcal{T}_2 \approx 0.065s$ for the fast network.

For the simulations, we consider a network of 4 Kuramoto oscillators with topology $\mathcal{G}$ given in Figure 4.5(b). For every edge in $\mathcal{G}$, we implement an individual PSCN channel. All channels have the same delay distribution ϕ but different realizations of the random server processing times. We use the following simulation parameters: $\omega_e = 10$ and $K = 10$. We compute the steady-state frequency $\overline{\omega}_1 = 5.79$ for the slow network and $\overline{\omega}_2 = 7.30$ for the fast network. In both cases, we have $\cos(\overline{\omega}_\upsilon \overline{\tau}_\upsilon) > 0, \upsilon = 1, 2$, and $\overline{z}_1 = 2.27$ as well as $\overline{z}_2 = 2.59$. Hence, these oscillators with distributed delay (4.22) synchronize for all values of $K \in \mathcal{K}$ and initial conditions satisfying $|\theta_i(\eta) - \overline{\omega}_\upsilon \eta - \tilde{\theta}_\upsilon| \leq \frac{\overline{z}_\upsilon}{2}$ for all $\eta \in [-\mathcal{T}, 0]$. Consider the following two initial conditions

$$\theta^{(1)}(\eta) = \begin{bmatrix} 0.25\pi + \omega_e(\eta + 1) \\ 0.11\pi + \omega_e(\eta + 1) \\ -0.09\pi + \omega_e(\eta + 1) \\ -0.12\pi + \omega_e(\eta + 1) \end{bmatrix}, \qquad \theta^{(2)}(\eta) = \begin{bmatrix} 0.95\pi + \omega_e(\eta + 1) \\ 0.2\pi + \omega_e(\eta + 1) \\ -0.43\pi + \omega_e(\eta + 1) \\ -0.15\pi + \omega_e(\eta + 1) \end{bmatrix},$$

for $\eta \in [-1, 0]$. Straightforward calculations show that $\theta^{(1)}$ satisfies $|\theta_i(\eta) - \overline{\omega}_\upsilon \eta - \tilde{\theta}_\upsilon| \leq \frac{\overline{z}_\upsilon}{2}$ for $\upsilon = 1, 2$, whereas $\theta^{(2)}$ does not satisfy this condition.

Exemplary simulation results of the Kuramoto oscillators with initial condition $\theta^{(1)}$ are presented in Figure 4.7. We see that the oscillators synchronize on both the fast and the slow network. In these simulations, the oscillators synchronize faster on the fast network. This can be seen more clearly in Figure 4.8 that depicts the standard deviation of the four phases θ_i over time. This correlation between average delay and convergence rate is not surprising: small delays lead to a faster convergence than large delays, see also Section 3.8.

Simulations for the second initial condition $\theta^{(2)}$ are depicted in Figure 4.9. The os-

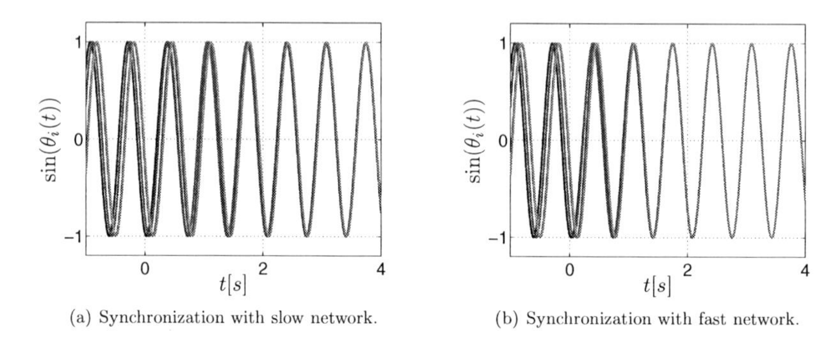

(a) Synchronization with slow network.

(b) Synchronization with fast network.

Figure 4.7: Simulation result for synchronization of Kuramoto oscillators (4.22) with initial condition $\theta^{(1)}$.

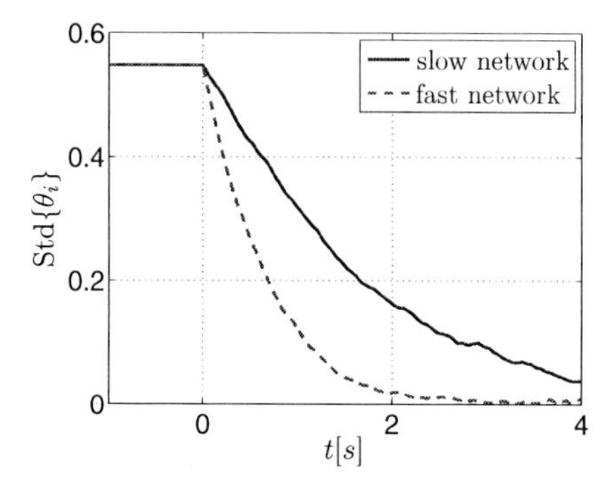

Figure 4.8: Standard deviation of phases θ_i of Kuramoto oscillators (4.22) over time.

cillators seem to synchronize on the fast network, see Figure 4.9(b), yet they do not synchronize on the slow network – at least not after $10s$, see Figure 4.9(a). Even more interesting, Figures 4.9(c) and 4.9(d) show the absolute phase of the oscillators, i.e. $\theta_i(t)$ instead of $\sin(\theta_i(t))$. There, we see that, even on the fast network, the phases do not reach a consensus as investigated in this thesis, i.e. $\lim_{t\to\infty}(\theta_i(t) - \theta_j(t)) \neq 0$ for all i, j. Note that the "synchronization" illustrated in Figures 4.9(a) and 4.9(b) only requires $\lim_{t\to\infty}(\theta_i(t) - \theta_j(t)) = 2k\pi$ for some $k \in \mathbb{Z}$ and for all i, j. The difference between the phases $\theta_i(t)$ in Figure 4.9(d) is 2π for $t \geq 4[s]$. This illustrates a fundamental restriction of this kind of vector space analysis for synchronization on limit cycles: For synchronization on limit cycles, it is irrelevant if the phases differ by an integer multiple of the length of the limit cycle. Analysis methods for linear vector spaces usually do not take this property into account. Nonetheless, we observe that (i) the developed methods

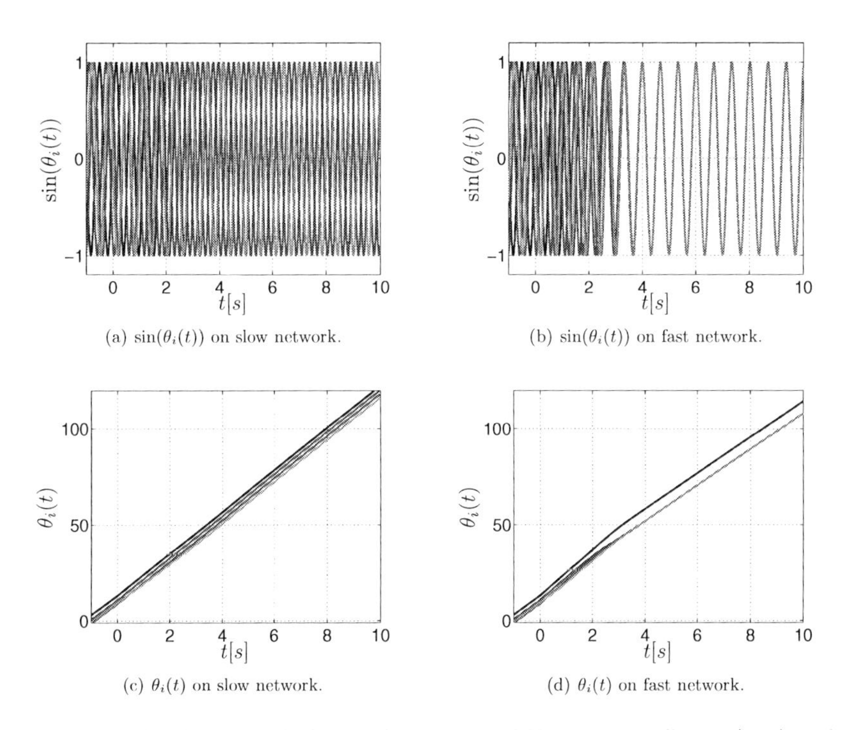

(a) $\sin(\theta_i(t))$ on slow network.

(b) $\sin(\theta_i(t))$ on fast network.

(c) $\theta_i(t)$ on slow network.

(d) $\theta_i(t)$ on fast network.

Figure 4.9: Simulation result for synchronization of Kuramoto oscillators (4.22) with initial condition $\theta^{(2)}$.

are useful to analyse the synchronization of Kuramoto oscillators with delays and (ii) the limitations of these methods to analyse synchronization originate in part from the structural differences between consensus in linear vector spaces and synchronization on manifolds, see also Sarlette (2009).

4.5 Summary

In this chapter, we showed that nonlinear RD1 MAS achieve rendezvous on directed, switching graphs independent of delay if the agent dynamics and the coupling functions satisfy certain conditions. These conditions require that the output of each agent follows the weighted average of the delayed outputs of its neighbours. These conditions are satisfied for single integrator MAS, the standard model in cooperative control. It has been shown in previous publications that this condition leads to rendezvous on directed switching graphs without delays (Lin, 2006; Moreau, 2004). The main contribution of this chapter is the extension of these results to MAS with heterogeneous delays preserving the generality of the conditions.

Chapter 5

Nonlinear Multi-Agent Systems with Relative Degree Two

In this chapter, we consider multi-agent systems consisting of a class of nonlinear agents with relative degree two. In contrast to the previous chapter, this MAS only achieves a rendezvous if the delays are sufficiently small. After presenting the MAS model in Section 5.1, we provide a scalable delay-dependent rendezvous condition for constant, time-varying, and distributed delays in Section 5.2. Therefore, we consider undirected graphs as in Chapter 3. Two application examples are provided in Section 5.3: Euler-Lagrange MAS and car-following models. For both applications, the developed results are used to derive rendezvous and flocking conditions. Finally, we compare in Section 5.4 the results of this chapter with the results of Section 3.6, where we studied linear MAS with relative degree two using the generalized Nyquist criterion. Moreover, we present simulation results of a linear MAS with relative degree two that communicates on a packet-switched communication network. The chapter is summarized in Section 5.5. Preliminary results of this chapter have been published in Münz et al. (2008b, 2010a,b).

5.1 Multi-Agent Systems with Relative Degree Two

In this chapter, we consider nonlinear, relative degree two MAS, for short *nonlinear RD2 MAS*, with the following agent dynamics illustrated in Figure 5.1

$$\Sigma_i : \begin{cases} \dot{x}_i(t) &= f_i(x_i(t), u_i(t)) \\ \tilde{y}_i(t) &= h_i(x_i(t)) \\ \dot{y}_i(t) &= \tilde{y}_i(t) \end{cases}, \qquad i \in \mathcal{N} = \{1, \ldots, N\}, \tag{5.1}$$

where $(x_i(t), y_i(t)) \in \mathbb{R}^{n_i} \times \mathbb{R}^m, u_i(t) \in \mathbb{R}^m, y_i(t) \in \mathbb{R}^m$ are the state, input, and output of agent i, respectively. We denote $\tilde{y}_i(t) \in \mathbb{R}^m$ *output derivative*. The functions $f_i : \mathbb{R}^{n_i} \times \mathbb{R}^m \to \mathbb{R}^{n_i}$ and $h_i : \mathbb{R}^{n_i} \to \mathbb{R}^m$ are Lipschitz continuous and satisfy $f_i(0,0) = h_i(0) = 0$. The initial condition of (5.1) interconnected with delays as defined below is $\varphi \in \mathcal{C}_{\sum n_i} = \mathcal{C}([-\mathcal{T}, 0], \mathbb{R}^{\sum n_i})$. We may think of (5.1) as a series interconnection of a relative degree one system with input u_i and output $\tilde{y}_i$ and an integrator with input $\tilde{y}_i$ and output y_i, see Figure 5.1. This particular structure with an integrator at the output implies that the output $y_i(t)$ approaches a constant value if the control input $u_i(t)$ is such that $\tilde{y}_i(t)$ goes to zero. The challenge is to design coupling functions such that all $y_i(t)$ end up with the same value.

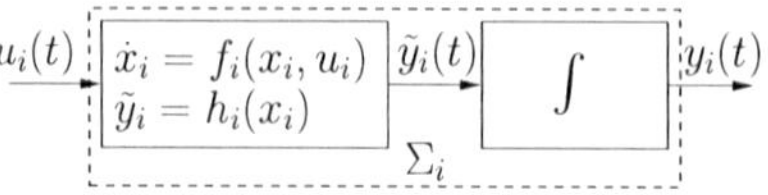

Figure 5.1: Block diagram of nonlinear RD2 agent dynamics Σ_i, see (5.1).

For illustration, we may think of $y_i, \tilde{y}_i$, and u_i as the position, velocity, and force input of a robot. Thereby, we see a fundamental distinction between the results in Chapter 4 and the results of this chapter. Newton's second law implies that the force input only affects the acceleration of the robots, i.e. the robots have relative degree one from force input to velocity output and relative degree two from force input to position output. Therefore, the results of Chapter 4 only apply if these robots have the cooperative control task to drive with the same velocity. If the robots have to achieve position rendezvous using position feedback alone, these robots constitute a RD2 MAS. In this case, the assumptions of Chapter 4 do not hold and new conditions for rendezvous and consensus are necessary.

Having the multi-robot application in mind, we assume the following: if no input, i.e. no external force, is applied to the robot, it slows down and eventually stops at any position. On the contrary, the robot does not stop as long as the external force does not vanish. The cooperative control task is to make all robots stop at the same position. More precisely, we assume the following:

Assumption 5.1. *The agents* (5.1) *satisfy the following:*

- *All agents are* strictly passive *from input u_i to the output derivative $\tilde{y}_i$, i.e. there exists a positive semidefinite, radially unbounded storage function $S_i(x_i)$ such that $\dot{S}_i(x_i) \leq -W_i(x_i) + u_i^T \tilde{y}_i$, where $W_i(x_i)$ is a positive definite function.*
- *There exist $\rho_i > 0$ and $\overline{h}_i > 0$ such that $W_i(x_i) \geq \rho_i \|x_i\|^2$ and $\|h_i(x_i)\|^2 \leq \overline{h}_i \|x_i\|^2$ for all x_i.*
- *The function f_i is such that $f_i(0, u_i) \neq 0$ for $u_i \neq 0$.*

The strict passivity assumption implies that the agents have relative degree one from u_i to $\tilde{y}_i$ and relative degree two from u_i to y_i. Moreover, strict passivity guarantees that a zero input drives the output derivative eventually to zero. The parameter ρ_i is a lower bound on the damping of the agent and $\overline{h}_i$ is an upper bound on the output function h_i. We will show in Section 5.3.1 how ρ_i and $\overline{h}_i$ are derived from a general Euler-Lagrange model. Both parameters ρ_i and $\overline{h}_i$ will play a pivotal role for the delay-dependent rendezvous conditions derived below. The condition that $f_i(0, u_i) \neq 0$ for $u_i \neq 0$ reflects the property that the robot does not stop for nonzero inputs.

As in Chapter 4, the nonlinear RD2 MAS aims at achieving rendezvous, i.e.

$$\begin{aligned} \lim_{t\to\infty} \|y_i(t) - y_j(t)\| &= 0, \\ \lim_{t\to\infty} \dot{y}_i(t) &= 0, \end{aligned} \tag{5.2}$$

for all $i, j \in \mathcal{N}$. In order to achieve this cooperation, we consider the same interconnection as in (4.4)

$$u_i(t) = -\sum_{j=1}^{N} \frac{a_{ij}}{d_i} k_{ij} \left(y_i(t) - \mathbb{T}_{ij}(y_{j,t}) \right), \tag{5.3}$$

where a_{ij} are the elements of the adjacency matrix of the underlying graph, $d_i = \sum_{j=1}^{N} a_{ij}$ is the degree of agent i, and $k_{ij} : \mathbb{R}^m \to \mathbb{R}^m$ are Lipschitz continuous, nonlinear coupling functions. The functions f_i, h_i, and k_{ij} are sufficiently smooth such that the closed loop system has piecewise continuously differentiable solutions, even for piecewise continuous time-varying delays, see Hale and Lunel (1993). As in Chapter 3, we consider here again undirected graphs, i.e. $a_{ij} = a_{ji}$. As in Chapter 4, we consider all three delay models for the delay operator $\mathbb{T}_{ij}(y_{j,t})$, i.e. constant, time-varying, or distributed delays.

We impose the following design condition on the coupling functions k_{ij}:

Design Condition 5.2. *The coupling functions $k_{ij} : \mathbb{R}^m \to \mathbb{R}^m$ are such that $k_{ij}(z) = \overline{k_{ij}}(\|z\|)\frac{z}{\|z\|}$ with nonlinear, continuously differentiable gain $\overline{k_{ij}} : \mathbb{R}_0^+ \to \mathbb{R}_0^+$ that satisfies $\overline{k_{ij}}(0) = 0$ and $\overline{k_{ij}}(\|z\|) > 0$ for all $\|z\| \neq 0$. Moreover, the coupling functions are symmetric, i.e. $\overline{k_{ij}}(\|z\|) = \overline{k_{ji}}(\|z\|)$, and there exist $\kappa_{ij} = \kappa_{ji} > 0$ such that $\max_{\eta > 0} \left\{ \frac{d\overline{k_{ij}}(\eta)}{d\eta}, \frac{\overline{k_{ij}}(\eta)}{\eta} \right\} \leq \kappa_{ij}$.*

Design Condition 5.2 is very similar to Design Condition 4.8. Basically, the input u_i is heading toward the delayed outputs of the neighbours of agent i. The bound κ_{ij} is a Lipschitz constant for the coupling function k_{ij} as will be shown in the proof of Theorem 5.3. Note in particular that $\frac{\overline{k_{ij}}(\eta)}{\eta} \leq \kappa_{ij}$ guarantees that the gain $\overline{k_{ij}}(\eta)$ converges "fast" to zero as η goes to zero. This implies also that k_{ij} is continuous at the origin and in particular $\overline{k_{ij}}(\|y_{ij}\|)\frac{y_{ij}}{\|y_{ji}\|} = 0$ if $\|y_{ji}\| = 0$.

We first provide a general condition for nonlinear RD2 MAS to achieve rendezvous in Section 5.2. These results are then applied to Euler-Lagrange systems and car-following models in Section 5.3.

5.2 Rendezvous on Constant Topologies

The following theorem states the main result of this chapter: a local delay-dependent rendezvous condition for nonlinear RD2 MAS.

Theorem 5.3 (Rendezvous in nonlinear RD2 MAS). *A MAS with agent dynamics* (5.1) *that satisfy Assumption 5.1 and with interconnection* (5.3) *where k_{ij} satisfy Design Condition 5.2 achieves rendezvous asymptotically on networks of arbitrary size $N \in \mathbb{N}$, with arbitrary connected undirected topologies, and arbitrary, bounded, heterogeneous con-*

stant, time-varying, or distributed delays if the coupling functions k_{ij} are such that

$$\kappa_{ij} < \frac{2\rho_i}{\overline{h}_i(\tau_{ij}+\tau_{ji})}, \quad \textit{for all } i,j \in \mathcal{N}, \quad \textit{for constant delays,} \tag{5.4a}$$

$$\kappa_{ij} < \frac{2\rho_i}{\overline{h}_i(\mathcal{T}_{ij}+\mathcal{T}_{ji})}, \quad \textit{for all } i,j \in \mathcal{N}, \quad \textit{for time-varying delays,} \tag{5.4b}$$

$$\kappa_{ij} < \frac{2\rho_i}{\overline{h}_i(\overline{\tau}_{ij}+\overline{\tau}_{ji})}, \quad \textit{for all } i,j \in \mathcal{N}. \quad \textit{for distributed delays,} \tag{5.4c}$$

where $\mathcal{T}_{ij}$ is the upper bound of the time-varying delay, i.e. $\tau_{ij}(t) \leq \mathcal{T}_{ij}$, and $\overline{\tau}_{ij} = \int_0^{\mathcal{T}_{ij}} \eta\phi_{ij}(\eta)d\eta$ is the average delay of the distributed delay.

Proof. Consider the Lyapunov-Krasovskii candidate $V_{\text{cd|td|dd}} = V_1 + V_2 + \tilde{V}_{\text{cd|td|dd}}$, where the indices are "cd" for constant delay, "td" for time-varying delay, and "dd" for distributed delay, with

$$V_1 = \sum_{i=1}^{N} d_i S_i(x_i) \tag{5.5}$$

$$V_2 = \frac{1}{2}\sum_{i=1}^{N}\sum_{j=1}^{N} a_{ij} \int_0^{\|y_{ij}\|} \overline{k_{ij}}(\eta)d\eta, \tag{5.6}$$

where S_i are the storage functions of the agents, see Assumption 5.1, and $y_{ij}(t) = y_i(t) - y_j(t)$. The last term $\tilde{V}_{\text{cd|td|dd}}$ is chosen depending on the type of delay, i.e.

$$\tilde{V}_{\text{cd}} = \frac{1}{2}\sum_{i=1}^{N}\sum_{j=1}^{N} a_{ij}\kappa_{ij} \int_0^{\tau_{ij}} \int_0^{\eta} \|\dot{y}_j(t+\xi)\|^2 d\xi d\eta \qquad \text{for constant delays,}$$

$$\tilde{V}_{\text{td}} = \frac{1}{2}\sum_{i=1}^{N}\sum_{j=1}^{N} a_{ij}\kappa_{ij} \int_0^{\mathcal{T}_{ij}} \int_0^{\eta} \|\dot{y}_j(t+\xi)\|^2 d\xi d\eta \qquad \text{for time-varying delays,}$$

$$\tilde{V}_{\text{dd}} = \frac{1}{2}\sum_{i=1}^{N}\sum_{j=1}^{N} a_{ij}\kappa_{ij} \int_0^{\mathcal{T}_{ij}} \int_0^{\eta} \int_0^{\xi} \phi_{ij}(\eta)\|\dot{y}_j(t-\chi)\|^2 d\chi d\xi d\eta, \quad \text{for distributed delays.}$$

The derivative of V_1 and V_2 along solutions of the MAS (5.1), (5.3) are

$$\begin{aligned}
\dot{V}_1 &\leq -\sum_{i=1}^{N} d_i W_i(x_i) + \sum_{i=1}^{N} d_i \dot{y}_i^T(t) u_i(t) \\
&\leq -\sum_{i=1}^{N} d_i \rho_i \|x_i\|^2 - \sum_{i=1}^{N}\sum_{j=1}^{N} a_{ij} \dot{y}_i^T(t) k_{ij}\left(y_i(t) - \mathbb{T}_{ij}(y_{j,t})\right), \\
\dot{V}_2 &= \frac{1}{2}\sum_{i=1}^{N}\sum_{j=1}^{N} a_{ij}\overline{k_{ij}}(\|y_{ij}\|)\frac{d\|y_{ij}\|}{dt} = \frac{1}{2}\sum_{i=1}^{N}\sum_{j=1}^{N} a_{ij}\overline{k_{ij}}(y_{ij})\dot{y}_{ij}^T\frac{y_{ij}}{\|y_{ij}\|} \\
&= \frac{1}{2}\sum_{i=1}^{N}\sum_{j=1}^{N}\left(a_{ij}\dot{y}_i^T k_{ij}(y_{ij}) - a_{ji}\dot{y}_i^T k_{ji}(y_{ji})\right) = \sum_{i=1}^{N}\sum_{j=1}^{N} a_{ij}\dot{y}_i^T k_{ij}(y_{ij}),
\end{aligned}$$

where we use the fact that the graph is undirected and Design Condition 5.2 in the last step. All together, we have

$$\dot{V}_1 + \dot{V}_2 \leq -\sum_{i=1}^{N} d_i\rho_i\|x_i\|^2 - \sum_{i=1}^{N}\sum_{j=1}^{N} a_{ij}\dot{y}_i^T(t)\left(k_{ij}(y_i(t) - \mathbb{T}_{ij}(y_{j,t})) - k_{ij}(y_{ij}(t))\right). \quad (5.7)$$

In order to simplify $\dot{V}_1 + \dot{V}_2$, we separate the delayed outputs $\mathbb{T}_{ij}(y_{j,t})$ from the undelayed outputs $y_i(t)$.

To this end, we show first that κ_{ij} determines a global Lipschitz bound for k_{ij}. Note that

$$\frac{\partial}{\partial z}\left(\frac{\overline{k_{ij}}(\|z\|)}{\|z\|}z\right) = \frac{\overline{k_{ij}}(\|z\|)}{\|z\|}I + \frac{\overline{k_{ij}}'(\|z\|)\|z\| - \overline{k_{ij}}(\|z\|)}{\|z\|^2}\frac{zz^T}{\|z\|} = \mathcal{M}(z),$$

for any $z \in \mathbb{R}^m$ and any continuously differentiable function $\overline{k_{ij}} : \mathbb{R}_0^+ \to \mathbb{R}_0^+$, where $\overline{k_{ij}}'(\|z\|) = \left.\frac{d\overline{k_{ij}}(\tilde{z})}{d\tilde{z}}\right|_{\tilde{z}=\|z\|}$. The matrix zz^T has rank one for $z \neq 0$. The only nonzero eigenvalue of zz^T is $\|z\|^2$ with corresponding eigenvector z. Hence, the eigenvalues of $\mathcal{M}(z)$ for given z are $\overline{k_{ij}}'(\|z\|)$ and $\frac{\overline{k_{ij}}(\|z\|)}{\|z\|}$ (with multiplicity $m-1$). Thus, Design Condition 5.2 guarantees that κ_{ij} is a Lipschitz bound for k_{ij}.

Now, we distinguish the three delay cases constant, time-varying, and distributed delay:

Constant delays First, we consider constant delays, i.e. $\mathbb{T}_{ij}(y_{j,t}) = y_j(t-\tau_{ij})$. We use $y_j(t-\tau_{ij}) = y_j(t) - \int_0^{\tau_{ij}} \dot{y}_j(t-\eta)d\eta$, Design Condition 5.2, and the triangle inequality to upper bound the last term in (5.7)

$$\begin{aligned}\dot{y}_i^T(t)\left(k_{ij}(y_i(t)-\mathbb{T}_{ij}(y_{j,t}))-k_{ij}(y_{ij})\right) &\leq \|\dot{y}_i(t)\|\,\|k_{ij}(y_i(t)-y_j(t-\tau_{ij}))-k_{ij}(y_{ij})\| \\ \leq \|\dot{y}_i(t)\|\kappa_{ij}\left\|\int_0^{\tau_{ij}}\dot{y}_j(t-\eta)d\eta\right\| &\leq \frac{1}{2}\kappa_{ij}\left(\tau_{ij}\|\dot{y}_i(t)\|^2+\frac{1}{\tau_{ij}}\left\|\int_0^{\tau_{ij}}\dot{y}_j(t-\eta)d\eta\right\|^2\right) \\ &\leq \frac{1}{2}\kappa_{ij}\left(\tau_{ij}\|\dot{y}_i(t)\|^2+\int_0^{\tau_{ij}}\|\dot{y}_j(t-\eta)\|^2d\eta\right),\end{aligned}$$

where we used Hölder's inequality (Hardy et al., 1952) in the last step to obtain

$$\begin{aligned}\left\|\int_0^{\tau_{ij}}\dot{y}_j(t-\eta)d\eta\right\|^2 &\leq \left(\int_0^{\tau_{ij}}\|\dot{y}_j(t-\eta)\|d\eta\right)^2 \\ &\leq \int_0^{\tau_{ij}}\|\dot{y}_j(t-\eta)\|^2d\eta\int_0^{\tau_{ij}}1d\eta=\tau_{ij}\int_0^{\tau_{ij}}\|\dot{y}_j(t-\eta)\|^2d\eta.\end{aligned}$$

Note that

$$\dot{\tilde{V}}_{\mathrm{cd}}=\frac{1}{2}\sum_{i=1}^N\sum_{j=1}^N a_{ij}\kappa_{ij}\left(\tau_{ij}\|\dot{y}_j(t)\|^2-\int_0^{\tau_{ij}}\|\dot{y}_j(t-\eta)\|^2d\eta\right).$$

Summarizing, we have

$$\begin{aligned}\dot{V}_{\mathrm{cd}}=\dot{V}_1+\dot{V}_2+\dot{\tilde{V}}_{\mathrm{cd}} &\leq -\sum_{i=1}^N\sum_{j=1}^N a_{ij}\rho_i\|x_i(t)\|^2+\frac{1}{2}\sum_{i=1}^N\sum_{j=1}^N a_{ij}\kappa_{ij}\tau_{ij}\left(\|\dot{y}_i(t)\|^2+\|\dot{y}_j(t)\|^2\right) \\ &\leq -\sum_{i=1}^N\sum_{j=1}^N a_{ij}\left(\rho_i-\frac{\tau_{ij}+\tau_{ji}}{2}\overline{h}_i\kappa_{ij}\right)\|x_i(t)\|^2,\end{aligned}$$

where we use Assumption 5.1 and the fact that $a_{ij}=a_{ji}$ and $\kappa_{ij}=\kappa_{ji}$ are symmetric. Hence, $\dot{V}_{\mathrm{cd}}\leq 0$ due to (5.4a).

Time-varying delays Next, we consider time-varying delays, i.e. $\mathbb{T}_{ij}(y_{j,t})=y_j(t-\tau_{ij}(t))$. We use $y_j(t-\tau_{ij}(t))=y_j(t)-\int_0^{\tau_{ij}(t)}\dot{y}_j(t-\eta)d\eta$, Design Condition 5.2, and the triangle inequality to obtain

$$\begin{aligned}\dot{y}_i^T(t)\left(k_{ij}(y_i(t)-y_j(t-\tau_{ij}(t)))-k_{ij}(y_{ij})\right) &\leq \|\dot{y}_i(t)\|\kappa_{ij}\left\|\int_0^{\tau_{ij}(t)}\dot{y}_j(t-\eta)d\eta\right\| \\ &\leq \frac{1}{2}\kappa_{ij}\left(\tau_{ij}(t)\|\dot{y}_i(t)\|^2+\int_0^{\tau_{ij}(t)}\|\dot{y}_j(t-\eta)\|^2d\eta\right),\end{aligned}$$

where we used again Hölder's inequality. Moreover, we have

$$\dot{\tilde{V}}_{\mathrm{td}}=\frac{1}{2}\sum_{i=1}^N\sum_{j=1}^N a_{ij}\kappa_{ij}\left(\mathcal{T}_{ij}\|\dot{y}_j(t)\|^2-\int_0^{\mathcal{T}_{ij}}\|\dot{y}_j(t-\eta)\|^2d\eta\right).$$

With $\tau_{ij}(t) \leq \mathcal{T}_{ij}$, we obtain

$$\dot{V}_{\text{td}} = \dot{V}_1 + \dot{V}_2 + \dot{\tilde{V}}_{\text{td}} \leq -\sum_{i=1}^{N}\sum_{j=1}^{N} a_{ij}\left(\rho_i - \frac{\mathcal{T}_{ij}+\mathcal{T}_{ji}}{2}\overline{h}_i\kappa_{ij}\right)\|x_i(t)\|^2.$$

Hence, $\dot{V}_{\text{td}} \leq 0$ due to (5.4b).

Distributed delays Finally, we consider distributed delays $\mathbb{T}_{ij}(y_{j,t}) = \int_0^{\mathcal{T}_{ij}} \phi_{ij}(\eta)y_j(t-\eta)d\eta$. We use $y_j(t-\eta) = y_j(t) - \int_0^{\eta}\dot{y}_j(t-\xi)d\xi$, $\phi_{ij}(\eta) \geq 0$, $\int_0^{\mathcal{T}_{ij}}\phi_{ij}(\eta)d\eta = 1$, $\overline{\tau}_{ij} = \int_0^{\mathcal{T}_{ij}}\eta\phi_{ij}(\eta)d\eta$, Design Condition 5.2, and the triangle inequality to obtain

$$\begin{aligned}
\dot{y}_i^T(t)&\left(k_{ij}\left(y_i(t) - \int_0^{\mathcal{T}_{ij}}\phi_{ij}(\eta)y_j(t-\eta)d\eta\right) - k_{ij}(y_{ij})\right)\\
&\leq \|\dot{y}_i(t)\|\kappa_{ij}\left\|y_j(t) - \int_0^{\mathcal{T}_{ij}}\phi_{ij}(\eta)y_j(t-\eta)d\eta\right\|\\
&= \kappa_{ij}\|\dot{y}_i(t)\|\left\|\int_0^{\mathcal{T}_{ij}}\int_0^{\eta}\phi_{ij}(\eta)\dot{y}_j(t-\xi)d\xi d\eta\right\|\\
&\leq \kappa_{ij}\int_0^{\mathcal{T}_{ij}}\phi_{ij}(\eta)\|\dot{y}_i(t)\|\left\|\int_0^{\eta}\dot{y}_j(t-\xi)d\xi\right\|d\eta\\
&\leq \frac{\kappa_{ij}}{2}\int_0^{\mathcal{T}_{ij}}\phi_{ij}(\eta)\left(\eta\|\dot{y}_i(t)\|^2 + \frac{1}{\eta}\left\|\int_0^{\eta}\dot{y}_j(t-\xi)d\xi\right\|^2\right)d\eta\\
&\leq \frac{\kappa_{ij}}{2}\left(\overline{\tau}_{ij}\|\dot{y}_i(t)\|^2 + \int_0^{\mathcal{T}_{ij}}\int_0^{\eta}\phi_{ij}(\eta)\|\dot{y}_j(t-\xi)\|^2 d\eta\right),
\end{aligned}$$

where we used again Hölder's inequality. Note that

$$\dot{\tilde{V}}_{\text{dd}} = \frac{1}{2}\sum_{i=1}^{N}\sum_{j=1}^{N} a_{ij}\kappa_{ij}\left(\overline{\tau}_{ij}\|\dot{y}_j(t)\|^2 - \int_0^{\mathcal{T}_{ij}}\int_0^{\eta}\phi_{ij}(\eta)\|\dot{y}_j(t+\xi)\|^2 d\xi d\eta\right).$$

Thus, we obtain again

$$\dot{V}_{\text{dd}} = \dot{V}_1 + \dot{V}_2 + \dot{\tilde{V}}_{\text{dd}} \leq -\sum_{i=1}^{N}\sum_{j=1}^{N} a_{ij}\left(\rho_i - \frac{\overline{\tau}_{ij}+\overline{\tau}_{ji}}{2}\overline{h}_i\kappa_{ij}\right)\|x_i(t)\|^2.$$

Hence, $\dot{V}_{\text{cd}} \leq 0$ due to (5.4c).

Barbalat's lemma and rendezvous Now, we use Barbalat's lemma, e.g. Khalil (2002), to conclude $\lim_{t\to\infty} x_i(t) = 0$ for all $i \in \mathcal{N}$. For notational convenience, denote $V = V_{\text{cd}}, V = V_{\text{td}}$, or $V = V_{\text{dd}}$ depending on the kind of delay. Consider the following limit

$$\lim_{t\to\infty} V(t) - V(0) = \int_0^{\infty}\dot{V}(\eta)d\eta \leq -\sum_{i=1}^{N}\sum_{j=1}^{N} a_{ij}\left(\rho_i - \frac{\tau_{ij}+\tau_{ji}}{2}\overline{h}_i\kappa_{ij}\right)\int_0^{\infty}\|x_i(\eta)\|^2 d\eta,$$

in the case of constant delays. Similar conditions are obtained for time-varying and distributed delays if we exchange τ_{ij}, τ_{ji} by $\mathcal{T}_{ij}, \mathcal{T}_{ji}$ or $\overline{\tau}_{ij}, \overline{\tau}_{ji}$, respectively. Note that $V \geq 0$ and V is non-increasing, i.e. $V(t) \leq V(0)$ for all $t \geq 0$. Therefore, $\lim_{t\to\infty} V(t) - V(0)$ is negative and finite. This implies that all integrals $\int_0^\infty \|x_i(\eta)\|^2 d\eta, i \in \mathcal{N}$, are finite. Therefore, we may conclude $\lim_{t\to\infty} x_i(t) = 0$ with Barbalat's lemma if $\|x_i(t)\|^2$ is uniformly continuous, what we prove next.

Note that $\|x_i(t)\|^2$ is uniformly continuous if $x_i(t)$ is piecewise continuously differentiable and $\dot{x}_i(t)$ is bounded for all $t \geq 0$. The functions f_i, h_i, and k_{ij} are assumed sufficiently smooth such that $x_i(t)$ is piecewise continuously differentiable, even for switching time-varying delays. It remains to show that $\dot{x}_i(t)$ is bounded for all $t \geq 0$. The state derivative $\dot{x}_i = f_i(x_i, u_i)$ depends on the state x_i and u_i. Thus, $\dot{x}_i$ is bounded if x_i and u_i are bounded because f_i is continuous. We show briefly that x_i and u_i are bounded. The state x_i is bounded because S_i and therefore V_1 is radially unbounded by Assumption 5.1 and because $V(t) \leq V(0)$. Since x_i is bounded, all $\tilde{y}_i = \dot{y}_i, i \in \mathcal{N}$, are bounded because h_i is continuous. Moreover, the construction of V_2 implies that y_{ij} is bounded because $V(t) \leq V(0)$. Thus, we know that

$$\|y_i(t) - \mathbb{T}_{ij}(y_{j,t})\| \leq \|y_{ij}\| + \left\| \int_0^{\tau_{ij}} \dot{y}_j(t-\eta) d\eta \right\|$$

is bounded for bounded delays $\mathcal{T}$. From this, we conclude that the input u_i is bounded because k_{ij} is continuous and because all $\|y_i(t) - \mathbb{T}_{ij}(y_{j,t})\|, i, j \in \mathcal{N}$, are bounded. In summary, x_i and u_i are bounded by the construction of the Lyapunov-Krasovskii functional. This guarantees that $\dot{x}_i$ is bounded. Therefore, $\|x_i\|^2$ is uniformly continuous, i.e. Barbalat's lemma applies to the present case. We conclude $\lim_{t\to\infty} x_i(t) = 0$ for all $i \in \mathcal{N}$.

Since h_i is continuous and $h_i(0) = 0$, $\lim_{t\to\infty} x_i(t) = 0$ implies $\lim_{t\to\infty} \tilde{y}_i(t) = 0$, i.e. the output derivative converges to zero. This implies that all outputs $y_i(t)$ converge to a constant, and therefore,

$$\lim_{t\to\infty} (y_i(t) - y_j(t)) = y_{ij}^*$$

for some $y_{ij}^* \in \mathbb{R}^m$. Now, we have to distinguish two cases: (i) $y_{ij}^* = 0$ for all i, j and (ii) $y_{ij}^* \neq 0$ for any i, j. The first case implies $\lim_{t\to\infty} u_i(t) = 0$ for all i and due to $f(0,0) = 0$ also $\lim_{t\to\infty} \dot{x}_i(t) = 0$ for all i, i.e. rendezvous is eventually reached. It remains to show that $y_{ij}^* \neq 0$ for any i, j is not possible. This is proven by contradiction. Assume $y_{ij}^* \neq 0$ for some i, j. Design Condition 5.2 implies that $\lim_{t\to\infty} u_i(t) = u_i^* \neq 0$ for at least one $i \in \mathcal{N}$, say I. Assumption 5.1 implies that $\lim_{t\to\infty} \dot{x}_I(t) = f_I(0, u_I^*) \neq 0$ because $\lim_{t\to\infty} x_i(t) = 0$ for all $i \in \mathcal{N}$ and because $f_i(0, u_i) \neq 0$ for $u_i \neq 0$. This is however a contradiction to $\lim_{t\to\infty} x_i(t) = 0$ for all $i \in \mathcal{N}$. In summary, $\lim_{t\to\infty}(y_i(t) - y_j(t)) = 0$ for all $i, j \in \mathcal{N}$. □

The proof of Theorem 5.3 shows that rendezvous is achieved even for non-identical systems with completely different dynamics and of different order n_i. The only common property is that all of them have relative degree two and satisfy Assumption 5.1. Another advantage of the presented condition is that the structure of the controller (5.3) allows for a completely decentralized and scalable controller design because the conditions are local. The upper bounds κ_{ij} depend only on the system parameters ρ_i and $\overline{h}_i$ of the corresponding agent but not on the dynamics of the other agents. Moreover, κ_{ij} depend on the delays between agent i and j, but it is independent of the delays in other parts of the network. More precisely, they depend on the round trip time $\tau_{ij} + \tau_{ji}$ for constant delays, the maximal round trip time $\mathcal{T}_{ij} + \mathcal{T}_{ji}$ for time-varying delays, and the average round trip time $\overline{\tau}_{ij} + \overline{\tau}_{ji}$ for distributed delays. The round trip time can be easily measured in real applications using acknowledgement packets even without synchronizing clocks. The assumption $\overline{k_{ij}}(\eta) = \overline{k_{ji}}(\eta)$ in Design Condition 5.2 requires a joint design of both coupling functions between agent i and j, i.e. k_{ij} and k_{ji}. Yet, this does not affect the scalability of Theorem 5.3 because the coupling functions k_{ij} to all neighbours $j \in \mathcal{N}_i$ of agent i are independent. A joint design of k_{ij} and k_{ji} can be easily achieved by choosing a generic function $\overline{k}(\eta)$ for all agents with only a single parameter κ_{ij} that is tuned locally, i.e. $\overline{k_{ij}}(\eta) = \kappa_{ij}\overline{k}(\eta)$. This decentralized design is particularly useful for large and heterogeneous MAS.

It is interesting to note that the delay-dependent conditions (5.4) depend on the damping of the agents, i.e. larger damping allows for larger delays. This is due to the fact that we consider static output coupling functions $k_{ij}(y_i(t) - \mathbb{T}_{ij}(y_{j,t}))$, see (5.3). If we consider more sophisticated feedbacks, e.g. incorporating a feedback of the agent's own velocity as $\tilde{k}_{ij}(\tilde{y}_i(t), y_i(t) - \mathbb{T}_{ij}(y_{j,t}))$ in (4.3), then the "damping" can be easily increased using the control law $\tilde{k}_{ij}(\tilde{y}_i, y_i(t) - \mathbb{T}_{ij}(y_{j,t})) = -\overline{\rho}_i\tilde{y}_i + k_{ij}(y_i(t) - \mathbb{T}_{ij}(y_{j,t}))$, i.e. rendezvous can be achieved for arbitrarily large but bounded delays with suitably chosen $\overline{\rho}_i$. We also emphasize again the connection between the upper bound κ_{ij} on the coupling functions $\overline{k_{ij}}(\cdot)$, the lower bound on the damping ρ_i, and the round trip delays $\mathcal{T}_{ij} + \mathcal{T}_{ji}$ as in Sections 3.5, 3.6, and 3.7: larger delays require smaller gains or larger damping.

Another important finding is the relation between time-varying and distributed delays in Theorem 5.3. In Michiels et al. (2005); Münz et al. (2009b), it has been shown that fast time-varying or random delays can also be modeled in a simplified way using distributed delays. Thereby, the delay kernel ϕ_{ij} describes the delay density of the time-varying delay. If this simplification is possible, then the distributed delay model is preferable because condition (5.4c) based on the average delay $\overline{\tau}_{ij}$ is less restrictive than condition (5.4b) based on the maximal delay $\mathcal{T}_{ij}$.

5.3 Applications

We have shown in Theorem 5.3 that heterogeneous MAS with delayed feedback interconnection satisfying Assumption 5.1 can achieve rendezvous if the coupling functions are properly tuned. In this section, we show two examples of MAS that satisfy Assumption 5.1. We start with a general Euler-Lagrange system in Subsection 5.3.1. Subsection 5.3.2 introduces a car-following model and shows how the results in Theorem 5.3 can be used to analyse the delay robustness for this model.

5.3.1 Rendezvous in Euler-Lagrange Systems

We consider again Euler-Lagrange systems as in Section 4.3.1, see for example Ortega et al. (1998). Now, we consider agents that exchange their position y_i in contrast to Section 4.3.1, where the agents exchanged their velocities $\dot{y}_i$. We show how Theorem 5.3 can be applied to Euler-Lagrange systems without an input-output linearization. Euler-Lagrange systems describe a broad class of nonlinear systems that model, for example, mechanical systems like robots with omnidirectional drives. We consider fully actuated Euler-Lagrange systems, where we assume that the gravitational forces are compensated by some local internal controller. Agent i's dynamics are thus given by

$$M_i(y_i)\ddot{y}_i + C_i(y_i, \dot{y}_i)\dot{y}_i = -\frac{\partial F_i(\dot{y}_i)}{\partial \dot{y}_i} + u_i, \qquad i \in \mathcal{N}, \tag{5.8}$$

where $y_i \in \mathbb{R}^m$ is the vector of generalized configuration coordinates of agent i and $u_i \in \mathbb{R}^m$ is the vector of generalized forces acting on system i (Ortega et al., 1998). The generalized inertia matrices $M_i(y_i) \in \mathbb{R}^{m \times m}$, the vector of centrifugal and Coriolis forces $C_i(y_i, \dot{y}_i)\dot{y}_i \in \mathbb{R}^m$, and friction modeled by the Rayleigh dissipation function $F(\dot{y}_i)$ satisfy the following:

Assumption 5.4. *The inertia matrices M_i are symmetric and positive definite. Moreover, there exists an $\mu_i > 0$ such that $M_i(y_i) \geq \mu_i I$ for all y_i. The matrices C_i are defined such that $\dot{M}_i - 2C_i$ is skew symmetric, i.e. $\dot{M}_i - 2C_i = -\left(\dot{M}_i - 2C_i\right)^T$. The Rayleigh dissipation function F_i is continuously differentiable and satisfies $\left.\frac{\partial F_i(\dot{y}_i)}{\partial \dot{y}_i}\right|_{\dot{y}_i=0} = 0$ and*

$$\dot{y}_i^T \frac{\partial F_i(\dot{y}_i)}{\partial \dot{y}_i} \geq \rho_i \|\dot{y}_i\|^2 \tag{5.9}$$

for some $\rho_i > 0$.

Assumption 5.4 is standard for fully-damped Lagrangian systems (Ortega et al., 1998). The parameter ρ_i corresponds to the parameter ρ_i in Assumption 5.1. The second parameter in Assumption 5.1 is $\overline{h}_i = 1$

The aim of the Euler-Lagrange MAS (5.8) is to achieve rendezvous (5.2) using the interconnection (5.3). The corresponding conditions are provided in the following corollary:

Corollary 5.5 (Rendezvous in Euler-Lagrange systems). *An Euler-Lagrange MAS with agent dynamics* (5.8) *that satisfies Assumption 5.4 and with interconnection* (5.3) *where k_{ij} satisfy Design Condition 5.2 achieves rendezvous asymptotically on networks of arbitrary size $N \in \mathbb{N}$, with arbitrary connected undirected topologies, and arbitrary, bounded, heterogeneous constant, time-varying, or distributed delays if the coupling functions k_{ij} are such that* (5.4) *holds.*

Proof. The proof is based on the same Lyapunov-Krasovskii functional as in Theorem 5.3 with $S_i = \dot{y}_i^T M_i(y_i)\dot{y}_i$. The derivative of V_1 given in (5.5) along solutions of the MAS is

$$
\begin{aligned}
\dot{V}_1 &= \sum_{i=1}^{N} d_i \left(\frac{1}{2}\dot{y}_i^T \dot{M}_i(y_i)\dot{y}_i + \dot{y}_i^T M_i(y_i)\ddot{y}_i \right) \\
&= \sum_{i=1}^{N} d_i \left(\frac{1}{2}\dot{y}_i^T \dot{M}_i(y_i)\dot{y}_i - \dot{y}_i^T \left(C_i(y_i, \dot{y}_i)\dot{y}_i + \frac{\partial F_i(\dot{y}_i)}{\partial \dot{y}_i} - u_i \right) \right) \\
&= \sum_{i=1}^{N} d_i \left(-\dot{y}_i^T \frac{\partial F_i(\dot{y}_i)}{\partial \dot{y}_i} + \dot{y}_i^T u_i + \frac{1}{4}\dot{y}_i^T \left(\dot{M}_i - 2C_i + \left(\dot{M}_i - 2C_i \right)^T \right) \dot{y}_i \right) \\
&\leq - \sum_{i=1}^{N} d_i \left(\rho_i \|\dot{y}_i\|^2 - \dot{y}_i^T u_i \right),
\end{aligned}
$$

where we used Assumption 5.4. Following the proof of Theorem 5.3, we obtain

$$
\dot{V}_{\mathrm{cd}} \leq - \sum_{i=1}^{N} \sum_{j=1}^{N} a_{ij} \left(\rho_i - \frac{\tau_{ij} + \tau_{ji}}{2} \kappa_{ij} \right) \|\dot{y}_i(t)\|^2,
$$

and similarly for time-varying and distributed delays. Hence, (5.4) guarantees $\dot{V}_{\mathrm{cd|td|dd}} \leq 0$.

Since $S_i \geq \mu_i \|\dot{y}_i\|^2$ and $\dot{V}_{\mathrm{cd|td|dd}} \leq 0$, we know that $\dot{y}_i$ is bounded and therefore u_i and $\ddot{y}_i$ are bounded. Therefore, $\dot{y}_i$ is uniformly continuous, i.e. $\lim_{t\to\infty} \dot{y}_i(t) = 0$. The remainder of the proof is the same as the proof of Theorem 5.3. Note that $\ddot{y}_i(t) \neq 0$ if $\dot{y}_i(t) = 0$ and $u_i(t) \neq 0$, see (5.8). □

Corollary 5.5 shows that the results in Theorem 5.3 can be readily applied to many fully-actuated electro-mechanical systems with relative degree two.

5.3.2 Flocking in Car-Following Models

As a second example, we investigate a car-following problem. The *optimal velocity model* (OVM) for car-following has been proposed in Bando et al. (1994), see also Helbing (2001) for an overview on car-following models. The OVM considers a line of cars on a motor way with car 1 on its head and car $i-1$ being in front of car i. The dynamics of car i are

$$
\ddot{y}_i(t) = \zeta \left(\dot{y}^* - \dot{y}_i(t) + \overline{\kappa} \tanh(\tilde{\kappa}(y_{i+1}(t) - y_i(t) - \vartheta)) \right), \tag{5.10}
$$

where $y_i(t) \in \mathbb{R}$ is the position of car i on the motor way, $\zeta > 0$ describes the sensitivity of the driver, $\dot{y}^*$ represents the desired velocity of all cars, e.g. the speed limit, and $\overline{\kappa} \tanh(\tilde{\kappa}(y_i(t) - y_{i-1}(t) - \vartheta))$ describes the attitude of the driver to accelerate and decelerate depending on the desired distance ϑ to the preceding car, the actual distance $y_i(t) - y_{i-1}(t)$, and scalar gains $\overline{\kappa} > 0, \tilde{\kappa} > 0$. Parameter values based on data from Japanese motorways are given for example in Bando et al. (1998) as $\zeta = 2.0[\frac{1}{s}]$, $\dot{y}^* = 15.3384[\frac{m}{s}]$, $\overline{\kappa} = 16.8[\frac{m}{s}]$, $\tilde{\kappa} = 0.086[\frac{1}{m}]$, and $\vartheta = 25[m]$.

Here, we extend this model in three ways: First, we allow for different dynamics for the different cars and drivers, i.e. $\zeta, \overline{\kappa}, \tilde{\kappa}$, and ϑ are replaced by $\zeta_i, \overline{\kappa}_{ij} = \overline{\kappa}_{ji}, \tilde{\kappa}_{ij} = \tilde{\kappa}_{ji}$, and $\vartheta_{i(i+1)}$, respectively, where $\vartheta_{i(i+1)} = -\vartheta_{(i+1)i} > 0$ denotes the distance between car i and the following car $i+1$. Second, we consider drivers that accelerate depending on one, two, or more cars in front of them and behind them, i.e. we introduce an undirected graph topology where $a_{ij} = a_{ji} > 0$ if driver i perceives car j and driver j perceives car i. The assumption of an undirected graph is in fact necessary in order to apply our method. Third, we take the reaction time of the drivers into account, i.e. they perceive the position of the preceding and the following cars only after a certain reaction time τ_{ij}, see also Orosz (2005) for a detailed discussion on car-following models with reaction delays. The resulting car-following model is

$$\ddot{y}_i(t) = \zeta_i \left(\dot{y}^* - \dot{y}_i(t) - \sum_{j=1}^{N} \frac{a_{ij}}{d_i} \overline{\kappa}_{ij} \tanh\left(\tilde{\kappa}_{ij}\left(y_i(t) - y_j(t-\tau_{ij}) - \dot{y}^*\tau_{ij} - \vartheta_{ij}\right)\right) \right), \quad (5.11)$$

where $\vartheta_{ij} = \sum_{l=i}^{j-1} \vartheta_{l(l+1)}$ if $i < j$ and $\vartheta_{ij} = \sum_{l=j}^{i-1} \vartheta_{(l+1)l}$ if $j < i$. Moreover, $\dot{y}^*\tau_{ij}$ describes the drivers ability to predict the current position of the other cars supposing they are driving with the desired velocity $\dot{y}^*$, i.e. $y_j(t) \approx y_j(t-\tau_{ij}) + \dot{y}^*\tau_{ij}$ if $\dot{y}_j(t-\eta) \approx \dot{y}^*$ for $\eta \in [\tau_{ij}, 0]$. Note that this simple prediction does not require to measure or estimate the velocity of the other cars.

Next, we transform the MAS in a moving coordinate system

$$\begin{bmatrix} \Delta y_i(t) \\ \Delta \dot{y}_i(t) \end{bmatrix} = \begin{bmatrix} y_i(t) - t\dot{y}^* + \sum_{l=0}^{i-1} \vartheta_{l(l+1)} \\ \dot{y}_i(t) - \dot{y}^* \end{bmatrix}, \quad (5.12)$$

where $\vartheta_{01} = 0$. With this transformation, we obtain

$$\Delta\ddot{y}_i(t) = -\zeta_i \left(\Delta\dot{y}_i(t) + \sum_{j=1}^{N} \frac{a_{ij}}{d_i} \overline{\kappa}_{ij} \tanh(\tilde{\kappa}_{ij}(\Delta y_i(t) - \Delta y_j(t-\tau_{ij}))) \right), \quad (5.13)$$

which we separate in the following form

$$\Delta\ddot{y}_i(t) = -\zeta_i \left(\Delta\dot{y}_i(t) - u_i(t)\right) \quad (5.14a)$$

$$u_i(t) = -\sum_{j=1}^{N} \frac{a_{ij}}{d_i} \overline{\kappa}_{ij} \tanh(\tilde{\kappa}_{ij}(\Delta y_i(t) - \Delta y_j(t-\tau_{ij}))) \quad (5.14b)$$

into agent dynamics and feedback interconnection.

We briefly show that (5.14a) is of the same class as (5.1) and satisfies Assumption 5.1. The state x_i in (5.1) is $x_i = \Delta\dot{y}_i$ for (5.14a). Consider the radially unbounded storage function $S_i(x_i) = \frac{1}{2\zeta_i}\Delta\dot{y}_i^2$. The derivative of this storage function along the solutions of (5.14a) is $\dot{S}_i(x_i) = -\Delta\dot{y}_i(t)^2 + \Delta\dot{y}_i(t)u_i(t)$ and we obtain $\rho_i = 1$ and $\overline{h}_i = 1$. Note also that $f_i(0, u_i) = \zeta_i u_i = 0$ implies $u_i = 0$.

Comparing the feedback (5.14b) to (5.3), we obtain the coupling function $k_{ij}(z) = \overline{\kappa}_{ij}\tanh(\tilde{\kappa}_{ij}z)$ with gain $k_{ij}(z) = \overline{\kappa}_{ij}\tanh(\tilde{\kappa}_{ij}|z|)$. Remember that $\overline{\kappa}_{ij} = \overline{\kappa}_{ji}$ and $\tilde{\kappa}_{ij} = \tilde{\kappa}_{ji}$, i.e. $\overline{k_{ij}} = \overline{k_{ji}}$. Hence, Design Condition 5.2 is satisfied with $\kappa_{ij} = \overline{\kappa}_{ij}\tilde{\kappa}_{ij}$.

We summarize the delay robustness of the car-following model (5.11) in the following Corollary, which follows directly from Theorem 5.3:

Corollary 5.6 (Flocking in car-following model (5.11)). *The cars with dynamics* (5.11) *achieve flocking asymptotically on the desired velocity and distances, i.e.*

$$\lim_{t\to\infty} \dot{y}_i(t) - \dot{y}^* = 0 \qquad\qquad \lim_{t\to\infty} y_i(t) - y_{i+1}(t) = \vartheta_{i(i+1)}, \tag{5.15}$$

for all $i \in \mathcal{N}$ for arbitrary many cars $N \in \mathbb{N}$, with arbitrary connected undirected topologies, and arbitrary, bounded, heterogeneous constant delays if

$$\overline{\kappa}_{ij}\tilde{\kappa}_{ij} < \frac{2}{\tau_{ij} + \tau_{ji}}, \qquad \textit{for all } i, j \in \mathcal{N}. \tag{5.16}$$

We have shown in Section 3.3 that MAS without self-delay cannot achieve flocking in general. This also applies to MAS with linear agents with relative degree two. In the car-following model considered here, flocking is possible because the desired velocity $\dot{y}^*$ appears explicitly in the coupling, i.e. the agents only have to agree on their position. In this case, it is possible to transform the MAS without self-delays in a moving coordinate system using the transformation (5.12) similar to the rotating coordinate system for Kuramoto oscillators in Section 4.4.

We remark that the conditions in Corollary 5.6 guarantee that the cars eventually reach the desired distances. However, there is no guarantee about the transient behavior, e.g. if the cars keep a security distance between each other during the transient to the desired distance. For the parameter values given in Bando et al. (1998), Corollary 5.6 guarantees flocking if the round trip time satisfies $\tau_{ij} + \tau_{ji} < 1.384[s]$. Similar results can be obtained for time-varying and distributed delays.

Summarizing, we have seen in this section that the delay-dependent rendezvous conditions (5.4) are useful to analyse practical applications.

5.4 Comparison with Generalized Nyquist Consensus Condition

In this section, we compare Theorem 5.3 to the corresponding results in Corollary 3.11 in Section 3.6 based on the generalized Nyquist criterion. We consider a simple but instructive linear MAS with relative degree two and linear coupling. For this MAS, we first determine the delay robustness using the results of Corollary 3.11. Then, we study the delay robustness based on Theorem 5.3 and compare these results. The section is completed by some simulation results.

5.4.1 Agent Dynamics

We consider a MAS consisting of robots with the aim to achieve a rendezvous in the plain. The robot dynamics are

$$\ddot{y}_i(t) + \rho \dot{y}_i(t) = u_i(t), \tag{5.17a}$$

where $y_i(t) \in \mathbb{R}^2$, $\dot{y}_i(t) \in \mathbb{R}^2$, and $u_i(t) \in \mathbb{R}^2$ are the position, velocity, and input force of robot i, respectively. The linear damping is modeled by the parameter ρ. The linear feedback law is

$$u_i = -\sum_{j=1}^{N} \kappa \frac{a_{ij}}{d_i} \left(y_i(t) - y_j(t - \tau_{ij}) \right), \tag{5.17b}$$

where $\kappa > 0$ is the coupling gain and $\tau_{ij} \leq \mathcal{T}$. The cooperative control task is

$$\begin{aligned} \lim_{t\to\infty} \|y_i(t) - y_j(t)\| &= 0, \\ \lim_{t\to\infty} \dot{y}_i(t) &= 0, \end{aligned} \tag{5.18}$$

for all $i, j \in \mathcal{N}$. The following subsections compare different approaches how to investigate the delay robustness of this cooperative control task, i.e. determine the maximal $\mathcal{T}$ such that rendezvous is guaranteed.

5.4.2 Comparison of Consensus Conditions

First, we determine the delay robustness based on the generalized Nyquist criterion. We only consider SISO agent dynamics in Chapter 3. However, the differential equations for both elements of y_i in (5.17) are decoupled. Therefore, the results in Section 3.6 also apply for this MAS. Corollary 3.11 states that (5.17) achieves rendezvous independent of delay if

$$\rho^2 \geq 2\kappa. \tag{5.19}$$

If $\rho^2 < 2\kappa$, this MAS achieves rendezvous if

$$\mathcal{T} < \frac{2}{\sqrt{2\kappa - \rho^2}} \arctan \frac{\rho}{\sqrt{2\kappa - \rho^2}}. \tag{5.20}$$

Remember from Section 3.6 that this delay bound is maximal, i.e. if the left and right side of (5.20) are equal, then there exist delays and topologies such that rendezvous is not reached.

Theorem 5.3 states that (5.17) achieves consensus if

$$\mathcal{T} \leq \frac{\rho}{\kappa}. \tag{5.21}$$

The resulting stability curves are depicted in Figure 5.2. The dash-dotted line indicates the delay-independent condition (5.19), i.e. for parameter values above this curve, the

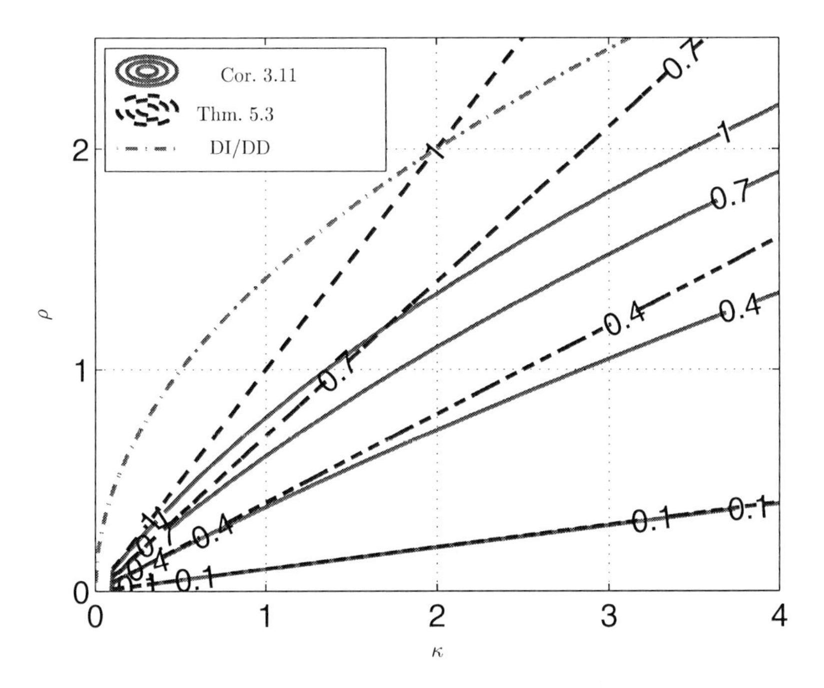

Figure 5.2: Contour plot of delay bound $\mathcal{T}$ (indicated as numbers on the lines) with respect to the coupling gain κ and the damping parameter ρ for MAS (5.17). For parameter triples $(\mathcal{T}, \kappa, \rho)$, Corollary 3.11 guarantees rendezvous if the point (κ, ρ) is above the solid line corresponding to $\mathcal{T}$. The solid lines are given by (5.20). Similarly, Theorem 5.3 guarantees rendezvous if the point (κ, ρ) is above the dashed line corresponding to $\mathcal{T}$. The dashed lines are given by (5.21). Finally, the parameter set (κ, ρ) where delay-independent rendezvous is guaranteed by Corollary 3.11 is above the dash-dotted line, indicated as boundary between delay-independent and delay-dependent (DI/DD) parameter sets according to (5.19). See Section 5.4.2 for more details.

MAS reaches a rendezvous independent of delay. The solid lines indicate the contours of the delay-dependent condition (5.20) based on the generalized Nyquist conditions. For example, if the the delay bound is $\mathcal{T} = 0.4$, any parameter combination above the solid 0.4-contour guarantees rendezvous. Finally, the dash-lines show the contours of the delay-dependent condition (5.21) based on Theorem 5.3. For small delays, condition (5.21) is in this example only slightly more conservative than the condition based on Corollary 3.11. This is even more surprising since condition (5.21) holds also for non-identical and nonlinear agent dynamics with time-varying delays whereas (5.20) is limited to linear MAS with identical agent dynamics and constant delays. This suggests that the derived conditions in Theorem 5.3 are suitable for a decentralized controller design with guaranteed robustness to delays.

5.4.3 Simulation Results

Finally, we illustrate the previous findings in a simulation. Therefore, we consider a MAS consisting of four robots with dynamics (5.17a) with $\rho = 0.1$. The agents exchange their position using a packet-switched communication network (PSCN) which introduces fast time-varying, randomly distributed packet delays. These delays can either be modeled as time-varying delays or as distributed delays with corresponding delay kernel. We illustrate in these simulations that distributed delay models may lead to more accurate delay bounds than time-varying delays.

Due to the PSCN, the linear feedback law is slightly modified as follows:

$$u_i = -\sum_{j=1}^{N} \kappa \frac{a_{ij}}{d_i} \left(y_i(t) - y_j(t - \tau_{ij}(t)) \right), \tag{5.22}$$

where $\tau_{ij}(t)$ describes the time-varying packet-delays. The corresponding distributed delay model is

$$u_i = -\sum_{j=1}^{N} \kappa \frac{a_{ij}}{d_i} \left(y_i(t) - \int_0^{\mathcal{T}_{ij}} \phi_{ij}(\eta) y_j(t - \eta) d\eta \right), \tag{5.23}$$

where ϕ_{ij} describes the delay density of the packet delays.

The simulations are performed using MATLAB® Simulink®. The PSCN is simulated as in Section 4.4.2 with the SimEvents® toolbox (MathWorks, 2006). We compare again the slower and the faster network from Section 4.4.2 with transmission time $T_{p1} = 0.05s$ and $T_{p2} = 0.025s$, respectively. The corresponding sampling times are $T_{s1} = 0.005s$ and $T_{s2} = 0.0025s$. The parameters of the gamma-distributed random process generating the server time are $\alpha = 5$ as well as $\beta_1 = 0.005$ and $\beta_2 = 0.0025$ for the slow and the fast network, respectively. The average delay is $\overline{\tau}_{ij} \approx 0.075s$ for the slow network and $\overline{\tau}_{ij} \approx 0.0375s$ for the fast network. Similarly, the maximal delay is $\mathcal{T}_{ij} \approx 0.13s$ and $\mathcal{T}_{ij} \approx 0.065s$, respectively.

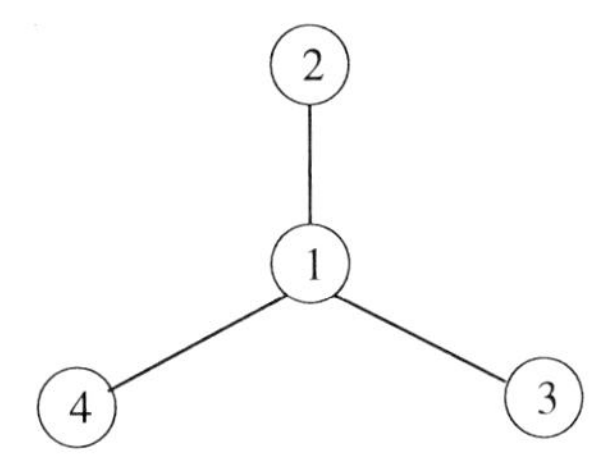

Figure 5.3: Topology of MAS with adjacency matrix (5.24).

We consider a MAS consisting of four agents with star topology, i.e. the adjacency matrix is

$$A = \begin{bmatrix} 0 & 1 & 1 & 1 \\ 1 & 0 & 0 & 0 \\ 1 & 0 & 0 & 0 \\ 1 & 0 & 0 & 0 \end{bmatrix}. \tag{5.24}$$

This topology is illustrated in Figure 5.3. It is suitable to evaluate the conservatism of Theorem 5.3 because (5.20) is exact in this case if the delays are homogeneous, see also Section 3.6. In total, we have 6 communication channels modeled with SimEvents®. Between every two connected agents, there are two individual PSCN channels, one from i to j and one from j to i. All channels have the same delay distribution ϕ_{ij} but different and independent realizations of the random delays $\tau_{ij}(t)$.

Before discussing the simulation results, we compute the delay robustness for both networks. Using the distributed delay model, κ depends on the average packet delays $\overline{\tau}_{ij}$. Therefore, we obtain $\kappa_{\mathrm{dd,s}} < 1.33$ for the slow network and $\kappa_{\mathrm{dd,f}} < 2.67$ for the fast network. Using the time-varying delay model, the corresponding values are $\kappa_{\mathrm{td,s}} < 0.77$ and $\kappa_{\mathrm{td,f}} < 1.54$, respectively. For our simulations, we choose $\kappa = 2$. Theorem 5.3 guarantees rendezvous of (5.17) with $\kappa = 2$ for the fast network but not for the slow network if the communication channels are modeled as distributed delays. If the channels are modeled as time-varying delays, Theorem 5.3 does not guarantee rendezvous for none of the two considered networks.

The initial condition for all simulations is

$$\begin{aligned} y(\eta) &= \begin{bmatrix} 5, & -2, & -3, & -5, & 2, & 2, & -1, & -2 \end{bmatrix}^T \\ \dot{y}(\eta) &= \begin{bmatrix} 2, & 1, & 0, & 5, & 1, & 3, & -1, & 5 \end{bmatrix}^T, \end{aligned}$$

for $\eta \in [-1, 0]s$. Simulation results for $t \in [0, 50]$ and $t \in [0, 100]$ are presented in Figures 5.4 and 5.5 for the slow and fast network, respectively. Each picture shows the position of the agents during $12.5s$ or $25s$, respectively. We see that the agents do not reach rendezvous on the slow network, in fact they spiral away. On the other hand, rendezvous is in fact achieved on the fast network, just as proven by Theorem 5.3.

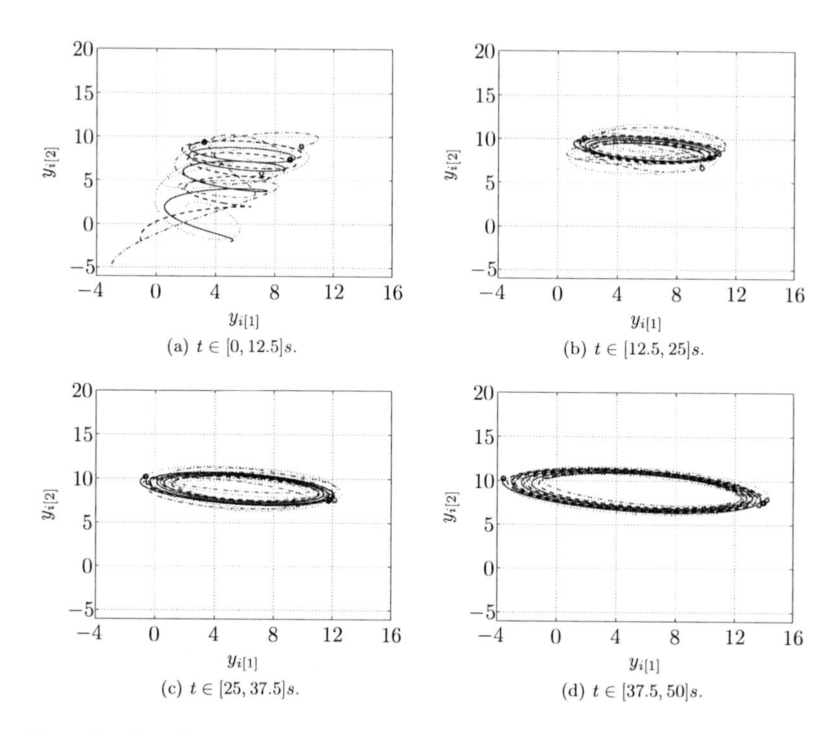

(a) $t \in [0, 12.5]s$.

(b) $t \in [12.5, 25]s$.

(c) $t \in [25, 37.5]s$.

(d) $t \in [37.5, 50]s$.

Figure 5.4: Simulation result of MAS (5.17a), (5.23) with $\rho = 0.1$ on the slow network with $\overline{\tau}_{ij} \approx 0.075s$.

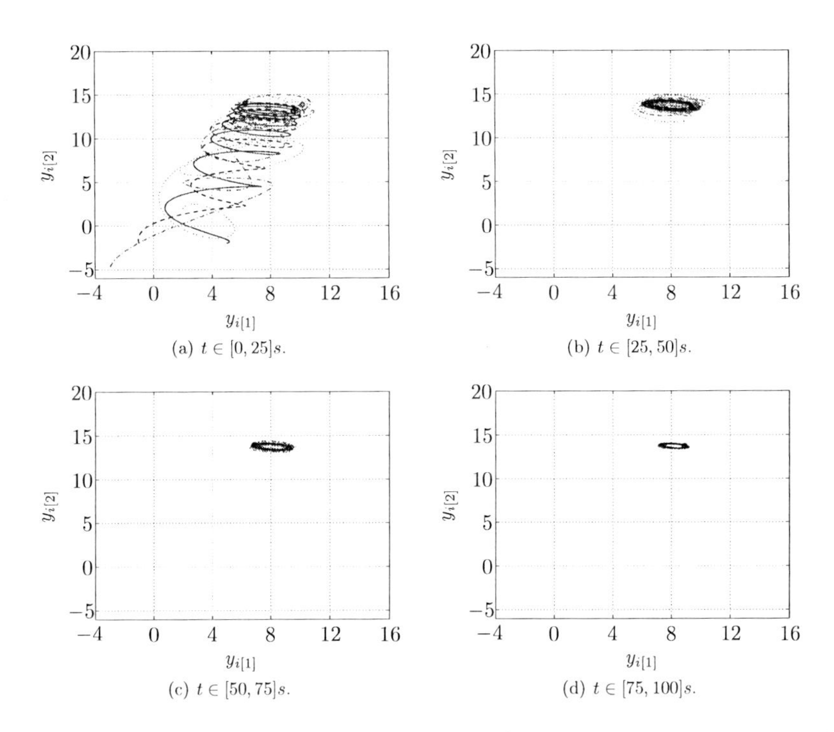

(a) $t \in [0, 25]s$.

(b) $t \in [25, 50]s$.

(c) $t \in [50, 75]s$.

(d) $t \in [75, 100]s$.

Figure 5.5: Simulation result of MAS (5.17a), (5.23) with $\rho = 0.1$ on the fast network with $\overline{\tau}_{ij} \approx 0.0375s$.

The simulations illustrate the findings of Section 5.4.2 and Section 3.6. In Section 3.6, we showed that Corollary 3.11 is quite accurate for topologies like the one considered here, where the adjacency matrix (5.24) has an eigenvalue at -1. In Section 5.4.2, we have seen that the conditions in Theorem 5.3 and in Corollary 3.11 are almost the same for MAS with small delays, see also Figure 5.2. In other words, the example is chosen such that the Conditions (5.4) should be quite accurate. The simulation results underline that Theorem 5.3 is indeed quite accurate in this example: Rendezvous is achieved on the fast network as guaranteed by Theorem 5.3 using distributed delays. For the slow network, the conditions in Theorem 5.3 are not satisfied and rendezvous is not achieved, even though Theorem 5.3 only provides sufficient conditions. Moreover, we see that distributed delays provide less conservative delay bounds than time-varying delays in the setup considered here. Yet, this only holds true if the time-varying delay is changing sufficiently fast in order to accurately model it using distributed delays.

5.5 Summary

We have developed local, scalable, delay-dependent rendezvous conditions for nonlinear RD2 MAS on undirected graphs. These conditions are particularly suitable for large networks of heterogeneous agents because they only depend on local dynamics and local delays but neither on the agent dynamics nor on delays in other parts of the network. The applications showed that the developed conditions can be used to analyse a broad range of MAS. Moreover, the comparison to the results in Section 3.6 revealed that the conditions of Theorem 5.3 are very close to the conditions based on the generalized Nyquist criterion for MAS with small damping and small delays. This is quite surprising because Theorem 5.3 extends the results of Section 3.6 to non-identical, nonlinear agent dynamics, nonlinear coupling functions, as well as time-varying and distributed delays.

Chapter 6

Conclusions

6.1 Summary

In order to achieve a cooperative behavior in multi-agent systems, agents have to exchange information over a network. These networks usually introduce some sort of delay. Even if these delays are small, they can be large enough to impede the cooperative behavior. This has been illustrated in a simulation example in Section 5.4.3. Therefore, the delay robustness of cooperative control is an important issue.

We presented in this thesis three different methods to analyse delay robustness in various cooperative control systems. In Chapter 3, we derived a set-valued consensus condition for general linear MAS, considering different feedback delay configurations. In Chapter 4, we showed that MAS composed of nonlinear agents with relative degree one achieve rendezvous independent of delay under the same conditions as in the undelayed case, even if the network's topology is switching. Finally, we derived local delay-dependent rendezvous conditions for MAS consisting of a class of nonlinear agents with relative degree two in Chapter 5. All results are scalable to large MAS with non-identical agents and heterogeneous delays. An overview of the different approaches is given in Table 1.1 in the Introduction.

One of the main contributions of this thesis is to prove that single integrator MAS without self-delay achieve consensus and rendezvous independent of delay, see Chapter 4. This holds in full generality for scalar and non-scalar single integrator agents with nonlinear coupling on directed, switching, uniformly quasi-strongly connected graphs and for constant, time-varying, and distributed delays under the same conditions on the coupling as in the undelayed case. Even if rendezvous and consensus are reached independent of delay, we expect intuitively that the convergence rate is delay-dependent. Therefore, we developed a delay-dependent condition for the convergence rate in linear single integrator MAS in Section 3.8. It has also been shown in Section 3.5 that consensus in single integrator MAS *with* self-delay is delay-dependent. That is, the delay robustness decreases if the agents' own output is delayed. These results are very important because single integrator MAS are by far the most studied model in the cooperative control literature.

We furthermore identified several classes of MAS where even very small delays can prevent a cooperative behavior. In general, cooperative control tasks can be corrupted by delays if the agent's own output is delayed, see Chapter 3 and in particular Section 3.4.3. For MAS without self-delay, we showed that consensus in relative degree two MAS is in general only achieved for sufficiently small delays and we derived delay-dependent conditions that guarantee consensus in Section 3.6. In Chapter 5, we developed local scalable delay-dependent rendezvous conditions for MAS consisting of a class of nonlinear agents with relative degree two (RD2). These conditions are suitable for a decentralized design of cooperative controllers for large groups of non-identical RD2 agents with guaranteed robustness to delays. The conservatism of this delay robustness has been investigated using conditions for linear MAS with RD2 agents in Section 5.4. Generalizing, we found out that larger delays usually can be compensated by smaller coupling gains.

In addition, we developed a unifying framework to analyse consensus in linear MAS with and without self-delay in Chapter 3. The resulting set-valued conditions allow for the first time to compare different MAS with and without delays. It turned out that MAS without self-delays are more robust to delays than MAS with self-delays for the most common agent dynamics, see Section 3.4.3.

These results underline that delay robustness is in fact an important issue for cooperative control. A wide spectrum of methods for a scalable delay robustness analysis of linear and nonlinear cooperative control systems have been developed in this thesis.

6.2 Outlook

The focus of this thesis is delay robustness in cooperative control. Some interesting areas of future work are outlined in the sequel. We first describe some nearby extensions of the presented results before indicating more distant research directions.

The method based on the generalized Nyquist criterion developed in Chapter 3 can also be applied to linear MAS with distributed delays. Therefore, we have to modify the sets Ω_r. Instead of considering the Laplace transform of constant delays $e^{-j\omega\tau_{ij}}$, we have to introduce the Laplace transform of the distributed delay kernel $\mathcal{L}(\phi_{ij})(s)$ in the proof of Lemma 3.5. Note that $|\mathcal{L}(\phi_{ij})(j\omega)| \leq 1$ because $\int_0^\infty \phi_{ij}(\eta)d\eta = 1$, i.e. the sets Ω_r remain subsets of the same discs as for the constant delay case. The characteristic angle $\omega\mathcal{T}$, that determines the size of the sets $\Omega_r(\omega\mathcal{T})$, depends in the case of distributed delays on the phase of the transfer function $\mathcal{L}(\phi_{ij})(j\omega)$. It would be interesting to investigate the connection between the delay kernels ϕ_{ij} and the phase of its transfer function in order to determine simple convex sets similar to the sets Ω_r. Moreover, the results in Chapter 3 can be easily extended to agent dynamics with computation or reaction delays.

Another interesting problem is the extensions of the convergence rate conditions in Section 3.8 to further performance criteria. For example, the robustness of sensor networks to measurement noise is investigated Olfati-Saber and Shamma (2005) for sensor networks without delays.

As already explained in Section 4.3.2, Chopra et al. (2008); Chopra and Spong (2006) investigate the rendezvous of Euler-Lagrange and passive systems, i.e. relative degree one systems, using sums of Lyapunov-Krasovskii functionals. Assuming that the underlying graph is balanced, a rendezvous controller is proposed that does not satisfy Design Condition 4.3. It would be interesting to study the relations between the results presented in Chapter 4, which are less restrictive on the network topology, and the results in Chopra et al. (2008); Chopra and Spong (2006), which are less restrictive on agent dynamics. This study may either lead to less restrictive conditions or a deeper insight why these results are complementary.

A future research area that is closely related to the problems considered in this thesis is distributed optimization. In distributed optimization problems, the agents have to find the global optimum based on local information. Thus, suitable conditions have to guarantee the asymptotic stability of this optimum. These decentralized optimization algorithms also rely on an exchange of information between the agents, which again introduces delays. Examples for these decentralized optimization problems, where the delay robustness has been investigated, are Internet congestion control (Papachristodoulou and Jadbabaie, 2010; Lestas and Vinnicombe, 2006, 2007a), power control in wireless networks (Charalambous et al., 2008; Charalambous and Ariba, 2009) or code-division multiple access (CDMA) (Fan and Arcak, 2006), and link level power control in optical networks (Stefanovic and Pavel, 2009a,b). However, there are still many open problems, e.g. because most of these results hold only for very simple linear dynamics. Therefore, it would be very interesting to extend the methods presented in this thesis to these problems.

Another very interesting field of future research are delay-coupled oscillator networks. Tyson (2005) and Novák and Tyson (2008) provide an overview on different delay-coupled oscillator networks in biochemical networks, such as the circadian oscillator (Wang et al., 2007; Scheper et al., 1999), genetic regulatory networks (Chen and Aihara, 2002; Wang et al., 2004, 2006; Bratsun et al., 2005), and neural networks (Castro et al., 1999). In biochemical networks, delays are often essential for a simplified modeling without partial differential equations and the delays can easily be in the range of several hours, e.g. Tyson (2005). A system theoretic viewpoint on general delay-coupled oscillator networks is given, for example, in Atay (2009); Atay and Karabacak (2006); Michiels and Nijmeijer (2009); Strogatz (1998). In oscillator networks, delays can play a pivotal role to enable oscillations or make oscillations more robust to parameter changes, e.g. Chen and Aihara (2002); Wang et al. (2004); Radde (2009); Novák and Tyson (2008). On the other hand, delays can destabilize limit cycles and lead to so-called amplitude death, e.g. Strogatz (1998); Atay (2009). Due to these opposite effects of delays, the analysis of delay-coupled oscillators is very challenging. So far, most of the results on oscillator networks only consider homogeneous delays, e.g. Chen and Aihara (2002); Wang et al. (2007); Bratsun et al. (2005); Tyson (2005); Novák and Tyson (2008); Castro et al. (1999); Atay and Karabacak (2006); Michiels and Nijmeijer (2009); Radde (2009). However, there are situations where heterogeneous delays appear and there are only very few results that consider multiple delays, e.g. Wang et al. (2004, 2006). The methods developed in this thesis can help to investigate delay-coupled oscillator networks with heterogeneous delays. On the one hand, they can be used directly to investigate a lower bound on the delays such that a stable system can turn into an oscillatory system. On the other hand, the methods could be developed further in order to investigate the robustness of oscillations for sufficiently large delays, e.g. introducing lower bounds on the delays.

A related problem to delay-coupled oscillators are cooperative control tasks for agents with multiple isolated equilibria. Interesting applications for this kind of consensus problems appear for instance in systems biology, where many phenomena like cell death are described using bistable systems. In this case, all cells should either die or survive. Yet, it is usually not desirable that half of the cells die and the rest survives. These MAS can only reach consensus at isolated points in contrast to the standard consensus problems where consensus can be reached anywhere in $\mathbb{R}^m$. This kind of consensus problem has so far only been investigated very rarely and without considering the influence of delays, see for example Wu (2009); Schmidt et al. (2010).

Finally, it would be very interesting to include even more complex properties of PSCN in the analysis of MAS. Quantization effects have already been investigated in Carli and Bullo (2009); Kashyap et al. (2007) and random packet dropouts are usually modeled as random graphs (Tahbaz-Salehi and Jadbabaie, 2008; Hatano and Mesbahi, 2005). Even more advanced models for packet-switched communication networks should take the correlation between the number of packets, the packet dropouts, and the delays into account.

Appendix A

Fundamentals of Retarded Functional Differential Equations

This appendix provides a brief overview on retarded functional differential equations (RFDE) and in particular on the stability analysis of RFDEs. A good introduction to the stability analysis of RFDEs with a focus on linear systems is given in Gu et al. (2003). For more details, the reader is referred to Hale and Lunel (1993); Michiels and Niculescu (2007); Niculescu (2001).

Let $\mathbb{R}^n$ denote the n-dimensional Euclidean space with the Euclidean norm $\|\cdot\|$. Let $\mathcal{C}_n([a,b],\mathbb{R}^n)$ denote the Banach space of continuous functions mapping the interval $[a,b] \subset \mathbb{R}$ into $\mathbb{R}^n$. For easier notation, we drop the argument of $\mathcal{C}_n$ if $a = -\mathcal{T}$ and $b = 0$ for a given *delay range* $\mathcal{T} > 0$, i.e. $\mathcal{C}_n = \mathcal{C}_n([-\mathcal{T},0],\mathbb{R}^n)$. The norm on $\mathcal{C}_n$ is defined as $\|\varphi\|_C = \sup_{\eta\in[-\mathcal{T},0]} \|\varphi(\eta)\|$. Let $\varrho \geq 0$ and $x \in \mathcal{C}_n([-\mathcal{T},\varrho],\mathbb{R}^n)$, then for any $t \subset [0,\varrho]$, we define a segment $x_t \in \mathcal{C}_n$ of x of length $\mathcal{T}$ such that $x_t(\eta) = x(t+\eta), \eta \in [-\mathcal{T},0]$.

Let $\mathfrak{D}$ be a subset of $\mathcal{C}$, $f : \mathfrak{D} \to \mathbb{R}^n$ a given function, and '' represent the right-hand derivative. Then, we call

$$\dot{x}(t) = f(x_t) \tag{A.1}$$

an autonomous *Retarded Functional Differential Equation* (RFDE) on $\mathfrak{D}$. Given $\varphi \in \mathfrak{D}$ and $\varrho > 0$, a function $x(\varphi) \in \mathcal{C}_n([-\mathcal{T},\varrho],\mathbb{R}^n)$ is said to be a solution to (A.1) with initial condition φ if $x_t(\varphi) \in \mathfrak{D}$, $x(\varphi)(t)$ satisfies (A.1) for $t \in [0,\varrho)$ and $x_0(\varphi) = \varphi$. Such a solution exists and is unique if f is continuous and $f(\varphi)$ is Lipschitz in each compact set in $\mathfrak{D}$, see Hale and Lunel (1993). Note that $x(\varphi)(t) \in \mathbb{R}^n$ is a vector, whereas $x_t(\varphi) \in \mathcal{C}_n$ is a segment of a vector valued function. We denote the value of the segment $x_t(\varphi)$ at time $\eta, \eta \in [-\mathcal{T},0]$, as $x_t(\varphi)(\eta) = x(\varphi)(t+\eta)$, where φ is the initial condition. For easier notation, we often drop the initial condition φ of x and x_t.

An element $\zeta \in \mathcal{C}$ is called a steady-state or equilibrium of (A.1) if $x_t(\zeta) = \zeta$ for all $t \geq 0$. In the following, we assume that $\zeta = 0$ is an equilibrium of (A.1). The stability of (A.1) around such a steady-state is defined in a way similar to the stability of nonlinear Ordinary Differential Equations (ODEs) using an ϵ-δ argument:

Definition A.1. *Suppose $f(0) = 0$. The solution $x_t = 0$ of (A.1) is said to be (uniformly)* stable *if for any $t_0 \in \mathbb{R}, \epsilon > 0$, there exists a $\delta = \delta(\epsilon) > 0$ such that $\|x_{t_0}\|_C < \delta$ implies $\|x_t\|_C < \epsilon$ for all $t \geq t_0$. It is (uniformly)* asymptotically stable *if it is (uniformly) stable and there exists a $\delta > 0$ such that for any $\epsilon > 0$, there exists a $T = T(\delta,\epsilon)$, such that $\|x_{t_0}\|_C < \delta$ implies $\|x_t\|_C < \epsilon$ for all $t \geq t_0 + T$ and all $t_0 \in \mathbb{R}$. It is* globally asymptotically stable *if it is asymptotically stable and* δ *can be an arbitrarily large, finite number.*

The stability of RFDEs can be analysed in the time-domain using Lyapunov-type arguments. There are two types of Lyapunov theorems for stability of equilibria of RFDEs, namely Lyapunov-Krasovskii and Lyapunov-Razumikhin. *Lyapunov-Krasovskii* is the natural extension of Lyapunov's theorem from ODEs to RFDEs. It is based on non-increasing Lyapunov-Krasovskii-functionals $V : \mathcal{C}_n \rightarrow \mathbb{R}$. The upper right-hand Dini derivative of V along the solution of (A.1) is

$$\dot{V}(x_t) = \limsup_{\Delta t \rightarrow 0^+} \frac{1}{\Delta t}(V(x_{t+\Delta t}) - V(x_t)). \tag{A.2}$$

Theorem A.2 (Lyapunov-Krasovskii Functional). *Suppose* $f : \mathfrak{D} \rightarrow \mathbb{R}^n$ *maps bounded sets in* $\mathfrak{D}$ *into bounded sets in* $\mathbb{R}^n$, *and* $v_1, v_2, v_3 : \mathbb{R}_0^+ \rightarrow \mathbb{R}_0^+$ *are continuous, non-decreasing functions, with* $v_1(\eta) > 0, v_2(\eta) > 0$ *for* $\eta > 0$ *and* $v_1(0) = v_2(0) = 0$. *If there exists a continuous functional* $V : \mathfrak{D} \rightarrow \mathbb{R}$ *such that*

$$\begin{aligned} v_1(\|x(t)\|) \leq V(x_t) &\leq v_2(\|x_t\|_C) \\ \dot{V}(x_t) &\leq -v_3(\|x(t)\|), \end{aligned}$$

then the solution $x_t = 0$ *of* (A.1) *is (uniformly) stable. If* $v_3(\eta) > 0$ *for* $\eta > 0$, *then it is (uniformly) asymptotically stable. If, in addition,* $v_1(\eta) \rightarrow \infty$ *as* $\eta \rightarrow \infty$, *then it is globally (uniformly) asymptotically stable.*

An alternative to Lyapunov-Krasovskii functionals are *Lyapunov-Razumikhin functions* $V(x(t)) : \mathbb{R} \rightarrow \mathbb{R}$, which use functions instead of functionals. As we only use Lyapunov-Krasovskii functionals in our proofs, we refer the interested reader to Gu et al. (2003); Hale and Lunel (1993) for more details on Lyapunov-Razumikhin functions.

If we consider linear time-invariant RFDEs, the stability can also be analysed in the Laplace domain. Consider the following RFDE with constant delays

$$\dot{x}(t) = \sum_{k=0}^{M} A_k x(t - \tau_k), \tag{A.3}$$

where $x(t) \in \mathbb{R}^n$, $A_k \in \mathbb{R}^{n \times n}$, and $\tau_k \in \mathbb{R}_0^+$. Without loss of generality, we assume $\tau_0 = 0$. The characteristic quasi-polynomial of (A.3) is

$$\Delta(s, \tau) = \det\left(sI - \sum_{k=0}^{M} A_k e^{-\tau_k s}\right) = \Delta_0(s) + \sum_{k=1}^{\tilde{M}} \Delta_k(s) e^{-\tilde{\tau}_k s}, \tag{A.4}$$

where $\tau = [\tau_1, \dots, \tau_N]^T$, Δ_k are polynomials in s, and $\tilde{\tau}_k$ are sums of the delays τ_j, see Gu et al. (2003). We recall the following lemma from Gu et al. (2003):

Lemma A.3. *System* (A.3) *is asymptotically stable if and only if all roots of* (A.4) *are in the open left half plane* $\mathbb{C}^-$.

Lemma A.3 suggests that the stability of (A.3) can be determined by computing the roots of (A.4). Note however that (A.4) has infinitely many roots. Hence, we need more sophisticated conditions that guarantee that all roots of (A.4) are in $\mathbb{C}^-$ without computing the roots explicitly. Most conditions in the literature rely on the continuity of the rightmost root of (A.4) with respect to τ, see for example Michiels and Niculescu (2007).

Lemma A.4. *All roots of $\Delta(s)$ are continuous with respect to the delays τ_k. Define the real part of rightmost root of Δ as $s_0 = \max\{\mathrm{Re}(s) : \Delta(s, \tau) = 0\}$. For any $\psi > s_0$, there exists a $\mathfrak{M} > 1$ such that any solution $x(t)$ of* (A.3) *with initial condition φ is bounded by*

$$\|x(t)\| \leq \mathfrak{M} e^{\psi t} \|\varphi\|_C.$$

The second part of this lemma will be used in our convergence rate analysis in Section 3.8.

Appendix B

Fundamentals of Graph Theory

This appendix provides a short introduction to graph theory as far as it is used in this thesis. The interested reader is referred to Godsil and Royle (2000) for more details.

A graph $\mathcal{G} = (\mathcal{V}, \mathcal{E})$ consists of a set of vertices or nodes $\mathcal{V} = \{v_i\}, i \in \mathcal{N} = \{1, \ldots, N\}$, and a set of edges or links $\mathcal{E} \subseteq \mathcal{V} \times \mathcal{V}$. In MAS, the agents are located at the vertices and the edges represent the interconnection links between the agents. If $v_i, v_j \in \mathcal{V}$ and $e_{ij} = (v_i, v_j) \in \mathcal{E}$, then there is an edge (a directed arrow) from node v_i to node v_j. In a MAS, this implies that agent j can receive data from agent i. A graph $\mathcal{G}$ is called *undirected* if $e_{ij} \in \mathcal{E}$ implies $e_{ji} \in \mathcal{E}$. Otherwise, the graph is called *directed*. Throughout this thesis, the considered network topology does not contain self-loops, i.e. $e_{ii} \notin \mathcal{E}$.

If $e_{ij} \in \mathcal{E}$, then v_j is a *parent* of v_i. In an undirected graph, we call v_i and v_j *neighbours* if $e_{ji} \in \mathcal{E}$. A *directed path* from v_i to v_j in a directed graph is a sequence of edges out of $\mathcal{E}$ that takes the following form $(v_i, v_{i_1}), (v_{i_1}, v_{i_2}), \ldots, (v_{i_p}, v_j)$. Accordingly, an *undirected path* from v_i to v_j in an undirected graph is a sequence of neighbouring vertices starting at v_i and ending at v_j. A *directed tree* is a directed graph where every vertex has exactly one parent except for one node, the so-called *root* v_R. In this case, there is a directed path from the root v_R to all other nodes of the directed tree.

A *subgraph* $(\tilde{\mathcal{V}}, \tilde{\mathcal{E}})$ of $\mathcal{G}$ is a graph with $\tilde{\mathcal{V}} \subseteq \mathcal{V}$ and $\tilde{\mathcal{E}} \subseteq \mathcal{E}$. If there exists a subgraph $(\mathcal{V}, \tilde{\mathcal{E}})$ of $\mathcal{G}$ that is a directed tree, then we say that $\mathcal{G}$ contains a *directed spanning tree*. Hence, a graph $\mathcal{G}$ contains a directed spanning tree if and only if it contains at least one *root*, i.e. one node with a directed path to all other vertices. For simplicity, we also say *spanning tree* when referring to a *directed spanning tree*.

A directed graph is *strongly connected* if there exists a directed path between any two vertices in the graph. The graph is called *quasi-strongly connected* if it contains a spanning tree. Note that quasi-strong connectivity is less restrictive than strong connectivity and in particular strong connectivity implies quasi-strong connectivity. Finally, an undirected graph is *connected* if there is an undirected path between any two vertices in the graph. Our results in Chapter 3 and 5 hold for MAS on connected undirected graphs. In Chapter 4, we investigate delay robustness of MAS on quasi-strongly connected directed graphs.

The *adjacency matrix* $A = [a_{ij}]$, $A \in \mathbb{R}^{N \times N}$, of the graph $\mathcal{G}$ is such that $a_{ij} > 0$ if $e_{ij} \in \mathcal{E}$ and $a_{ij} = 0$ if $e_{ij} \notin \mathcal{E}$. The graph is called *weighted* if a_{ij} take any positive real value instead of $a_{ij} \in \{0, 1\}$. If a weighted graph is undirected, then its weights are symmetric, i.e. $a_{ij} = a_{ji}$ and $A = A^T$. All results in this thesis hold for weighted graphs. The *in-degree* of vertex v_i is defined as $d_i = \sum_{j=1}^{N} a_{ij}$. The *degree matrix* of $\mathcal{G}$

is thus defined as $D = \text{diag}(d_i)$ and the *Laplacian matrix* is $L = D - A$. We also define the *normalized Laplacian matrix* $\overline{L} = D^{-1}L = I - D^{-1}A$.

Denote the eigenvalues of Laplacian matrix L as $\lambda_i, i \in \mathcal{N}$. If the graph is undirected, then all λ_i are real because L is symmetric. In this case, we assume without loss of generality $\lambda_1 \leq \lambda_2 \leq \ldots \leq \lambda_N$. If the graph is connected, then $\lambda_1 = 0$ and $\lambda_2 > 0$. The eigenvector of L corresponding to λ_1 is $\mathbf{1} = [1, 1, \ldots, 1]^T$. The second smallest eigenvalue λ_2 is called *algebraic connectivity* of the graph. It is a measure of how well the graph is connected. In Section 3.8, the second smallest eigenvalue of $\overline{L}$ will play a pivotal role for the convergence rate analysis. Note that all eigenvalues of the normalized Laplacian matrix $\overline{L}$ are also real if the graph is undirected, because $D^{-1}L$ has the same eigenvalues as $D^{-\frac{1}{2}}LD^{-\frac{1}{2}}$, see the proof of Lemma 3.5.

In Chapter 4, we also consider switching graphs $\mathcal{G} : \mathbb{R} \to \mathfrak{G}$ where $\mathfrak{G} = \{\mathcal{G}_p\}, p \in \mathcal{P} = \{1, \ldots, P\}$, is a finite set of P different directed graphs $\mathcal{G}_p = (\mathcal{V}, \mathcal{E}_p)$ with identical node set $\mathcal{V}$ but different edge set $\mathcal{E}_p$ and corresponding adjacency matrix $A_p \in \mathbb{R}^{N \times N}$. The edge set $\mathcal{E}(t) = \mathcal{E}_p$ and the adjacency matrix $A(t) = A_p$ at time t corresponds to the graph $\mathcal{G}(t) = \mathcal{G}_p$. The function $\mathcal{G}$ is piecewise constant from the right and we denote the time instances where $\mathcal{G}$ switches $t_\varsigma > t_{\varsigma-1}, \varsigma = 1, 2, \ldots$. The switching graph $\mathcal{G}$ switches with a dwell-time $h_{DW} > 0$, i.e. $t_\varsigma - t_{\varsigma-1} \geq h_{DW}$. This guarantees that the switching graph is non-chattering. The union graph over an interval $[t_1, t_2]$ is $\mathcal{G}([t_1, t_2]) = (\mathcal{V}, \bigcup_{t \in [t_1, t_2]} \mathcal{E}(t))$. A switching graph is *uniformly quasi-strongly connected* if there exists a $\mathfrak{T} > 0$ such that, for all $t \geq 0$, the union graph $\mathcal{G}([t, t + \mathfrak{T}])$ is quasi-strongly connected.

Appendix C

Technical Proofs

C.1 Proof of Corollary 3.9

Consider (3.32) and note that $\frac{1}{2+H_i(j\omega)} = \frac{1}{2+\frac{j\omega}{K_i}} < \frac{1}{2}$ for all $i \in \mathcal{N}$ and all $\omega \neq 0$. Moreover, all $z \in \Omega_1(\omega\mathcal{T})$ satisfy $|z| \leq 2$ for any $\omega\mathcal{T}$. Hence, (3.32) holds.

C.2 Proof of Corollary 3.10

We show first that (3.36) implies $2 + \frac{j\omega}{\tilde{K}} \notin \Omega_3(\omega\mathcal{T})$ for all $\tilde{K} \in (0, K]$ and all $\omega \neq 0$. Then, we prove in a second step that $2 + \frac{j\omega}{\tilde{K}} \notin \Omega_3(\omega\mathcal{T})$ for all $\tilde{K} \in (0, K]$ implies (3.32).

Consider Figure C.1(a) which shows $\Omega_3(\omega\mathcal{T}), \omega\mathcal{T} > 0$, and $2 + \frac{j\omega}{\tilde{K}}$ as dash-dotted line. We will show that (3.36) implies $2 + \frac{j\omega}{\tilde{K}} \notin \Omega_3(\omega\mathcal{T})$ for all $\tilde{K} \in (0, K]$ and all $\omega \neq 0$. Since the real part of $2 + \frac{j\omega}{\tilde{K}}$ is constant, we can restrict our analysis to the imaginary part, i.e. $\frac{j\omega}{\tilde{K}} \notin \{z \in \mathbb{C} : z \in \Omega_3(\omega\mathcal{T}), \mathrm{Re}(z) = 2\}$. Moreover, we only consider $\omega > 0$. Exactly the same conditions are obtained for $\omega < 0$ because both $2 + \frac{j\omega}{\tilde{K}}$ and $\Omega_3(\omega\mathcal{T})$ are symmetric to the imaginary axis for positive and negative ω. Note that $\Omega_3(\omega\mathcal{T}), \omega\mathcal{T} > 0$, always contains $z \in \mathbb{C}$ with $\mathrm{Re}(z) = 2$ and $\mathrm{Im}(z) > 0$ for $\omega > 0$. The imaginary part of these z is bounded by

$$\begin{aligned}\mathrm{Im}(z) &\leq \sin(\omega\mathcal{T}) + (1 - \cos(\omega\mathcal{T}))\tan(\frac{\omega\mathcal{T}}{2}) \\ &= \sin(\omega\mathcal{T})\left(1 + \frac{1 - \cos\omega\mathcal{T}}{1 + \cos\omega\mathcal{T}}\right),\end{aligned}$$

for all $\omega\mathcal{T} \in (0, \frac{\pi}{2}]$, see Figure C.1(a). We have to show that $\frac{\omega}{\tilde{K}} > \sin(\omega\mathcal{T})\left(1 + \frac{1-\cos\omega\mathcal{T}}{1+\cos\omega\mathcal{T}}\right)$ for all $\omega\mathcal{T} \in (0, \frac{\pi}{2}]$ and all $\tilde{K} \in (0, K]$. We reformulate this inequality as

$$\frac{1}{\tilde{K}\mathcal{T}} \geq \frac{1}{K\mathcal{T}} > \frac{\sin(\omega\mathcal{T})}{\omega\mathcal{T}}\left(1 + \frac{1 - \cos\omega\mathcal{T}}{1 + \cos\omega\mathcal{T}}\right) = f(\omega\mathcal{T}).$$

The derivative of f with respect to $\omega\mathcal{T}$ is

$$\frac{df(\omega\mathcal{T})}{d\omega\mathcal{T}} = 2\frac{\omega\mathcal{T} - \sin(\omega\mathcal{T})}{(\omega\mathcal{T})^2(1 + \cos(\omega\mathcal{T}))},$$

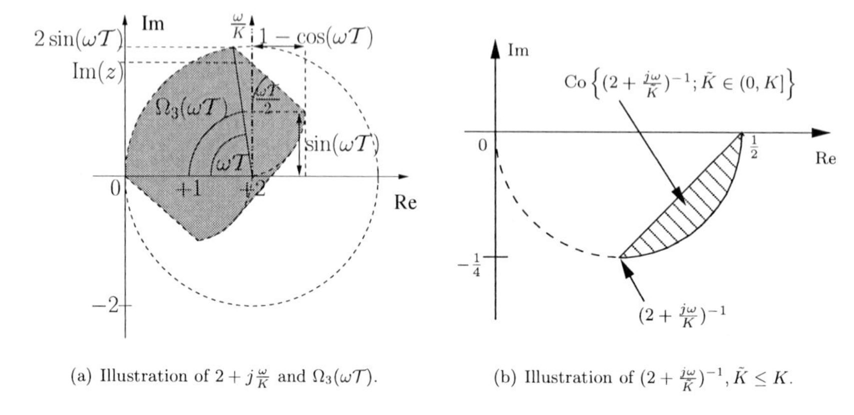

(a) Illustration of $2+j\frac{\omega}{K}$ and $\Omega_3(\omega\mathcal{T})$.

(b) Illustration of $(2+\frac{j\omega}{\tilde{K}})^{-1}, \tilde{K}\leq K$.

Figure C.1: Illustration for proof of Corollary 3.10.

i.e. f is strictly increasing for $\omega\mathcal{T}\in(0,\frac{\pi}{2}]$. Hence, $\frac{1}{K\mathcal{T}} > f(\frac{\pi}{2}) = \frac{4}{\pi}$, i.e. (3.36), guarantees $2+\frac{j\omega}{\tilde{K}} \notin \Omega_3(\omega\mathcal{T})$ for all $\tilde{K}\in(0,K]$ and all $\omega\neq 0$.

Now, we show that $-1\notin -(2+\frac{j\omega}{\tilde{K}})^{-1}\Omega_r(\omega\mathcal{T})$, i.e. $2+\frac{j\omega}{\tilde{K}}\notin\Omega_3(\omega\mathcal{T})$, for all $\tilde{K}\in(0,K]$ and all $\omega\neq 0$ implies (3.32). If $-1\notin -(2+\frac{j\omega}{\tilde{K}})^{-1}\Omega_r(\omega\mathcal{T})$ holds, then we have $-1\notin -\eta(2+\frac{j\omega}{\tilde{K}})^{-1}\Omega_r(\omega\mathcal{T})$ for any $\eta\in[0,1]$ because Ω_r is convex and contains the origin. Moreover, consider any $z\in\mathrm{Co}\left\{(2+\frac{j\omega}{K_i})^{-1}; i\in\mathcal{N}\right\}$ for any $K_i\in(0,K]$, and note that $z=\eta(2+\frac{j\omega}{\tilde{K}})^{-1}$ for appropriately chosen $\eta\in[0,1]$ and $\tilde{K}\in(0,K]$, see Figure C.1(b). Hence, (3.36) implies that $-1\notin -\mathrm{Co}\left\{(2+\frac{j\omega}{K_i})^{-1}\Omega_r(\omega\mathcal{T}); i\in\mathcal{N}\right\}$, i.e. (3.32), holds for $\Omega_r=\Omega_3$. Since $\Omega_4(\omega\mathcal{T})\subseteq\Omega_2(\omega\mathcal{T})\subseteq\Omega_3(\omega\mathcal{T})$, (3.32) holds for $r=2,3,4$.

C.3 Proof of Corollary 3.11

We apply Corollary 3.8 to (3.37) and see that

$$2+H^{-1}(j\omega) = 2-\frac{\omega^2}{K}+j\frac{\omega\rho}{K}.$$

This transfer function is illustrated in Figure C.2 for $K=1$ and $\rho\in\{1,1.5\}$. In order to reach consensus independent of delay, we have to guarantee that this transfer function neither touches nor enters the disc with center 1 and radius 1 for $\omega\neq 0$, i.e. the disc containing Ω_1, see the blue dashed line in Figure C.2. In order to simplify the analysis, we shift $2+H^{-1}(j\omega)$ and Ω_1 by 1 to the left. Then, $\Omega_1(\omega\mathcal{T})-1$ is a subset of the unit

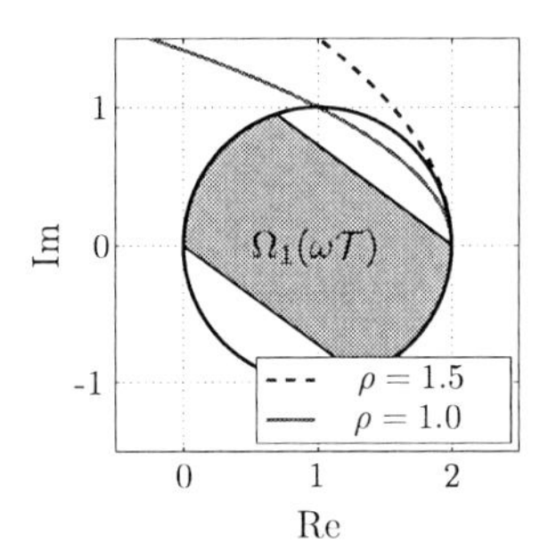

Figure C.2: Illustration of $\Omega_1(\omega\mathcal{T})$, (3.23a), for $\omega = 2.5$ and $\mathcal{T} = 0.5$ and $2 + H^{-1}(j\omega)$ with $H(s)$ given in (3.37) with $K = 1$ for $\rho = 1$ (red solid line) and $\rho = 1.5$ (blue dashed line), respectively.

disc for all ω and $1 + H^{-1}(j\omega)$ may not enter this unit disc. Thus, we have

$$\begin{aligned} \left|1 + H^{-1}(j\omega)\right| &> 1 \\ \left(1 - \frac{\omega^2}{K}\right)^2 + \left(\frac{\omega\rho}{K}\right)^2 &> 1 \\ \omega^2\left(\omega^2 + \rho^2 - 2K\right) &> 0, \end{aligned}$$

which is true for all $\omega \neq 0$ if (3.38) holds.

If $2 + H^{-1}(j\omega)$ enters the disc, consensus is still reached for sufficiently small delays, see the red solid line in Figure C.2. Note that $2 + H^{-1}(j\omega)$ is concave and the imaginary part is strictly increasing for $\omega \geq 0$. Thus, we only need to consider the point where $2 + H^{-1}(j\omega)$ leaves the disc because Ω_1 is convex. The frequency ω where it leaves the disc is obtained by solving $\left|1 + H^{-1}(j\omega)\right| = 1$, which gives $\omega = \sqrt{2K - \rho^2}$. Moreover, we have $H^{-1}\left(j\sqrt{2K - \rho^2}\right) = -\frac{2K-\rho^2}{K} + j\frac{\rho}{K}\sqrt{2K - \rho^2}$. Now, we shift $2 + H^{-1}(j\omega)$ and Ω_1 by 2 to the left and require that the phase of $H^{-1}\left(j\sqrt{2K - \rho^2}\right)$ is smaller than $\pi - \frac{\omega\mathcal{T}}{2}$ for $\omega = \sqrt{2K - \rho^2}$, see also Figure 3.4(a), i.e.

$$\begin{aligned} \pi - \arctan\frac{\rho}{\sqrt{2K - \rho^2}} &< \pi - \frac{1}{2}\sqrt{2K - \rho^2}\mathcal{T} \\ \arctan\frac{\rho}{\sqrt{2K - \rho^2}} &> \frac{1}{2}\sqrt{2K - \rho^2}\mathcal{T} \end{aligned}$$

which holds due to (3.39).

C.4 Proof of Corollary 3.12

We use again Corollary 3.8 to analyse (3.37) with different self-delay (3.3c). Consensus for (3.37) with identical self-delay (3.3b) holds under the same conditions because (3.3c) includes (3.3b) as special case.

This proof uses a slightly conservative approach. If we consider Figure C.1(a), we see that the imaginary part of all $z \in \Omega_3(\omega\mathcal{T})$ satisfies $\mathrm{Im}(z) \leq 2\sin\omega\mathcal{T}$. Thus, we have $2 + H^{-1}(j\omega) \notin \Omega_3(\omega\mathcal{T})$ if

$$\mathrm{Im}(2 + H^{-1}(j\omega)) = \frac{\rho}{K}\omega > 2\sin\omega\mathcal{T}, \qquad \text{for all } \omega \in (0, \frac{\pi}{2\mathcal{T}}].$$

This is satisfied if (3.41) holds because $\eta > \sin\eta, \forall\eta > 0$.

C.5 Proof of Theorem 3.17

First, we revisit the field of values of the normalized adjacency matrix for $s = -\psi + j\omega$. Following the proof of Lemma 3.5, we have

$$F(2I - \Gamma_1(-\psi + j\omega)) \subseteq \tilde{\Omega}_1(\omega\mathcal{T}) = \mathrm{Co}\left\{1 - e^{\psi\mathcal{T}}e^{-j\chi}, 1 + e^{\psi\mathcal{T}}e^{-j\chi} : \chi \in [0, \omega\mathcal{T}]\right\}.$$

Hence, the return ratio satisfies

$$\sigma\left(G_1(-\psi + j\omega)\right) \subseteq -\frac{1}{2 + H^{-1}(-\psi + j\omega)}\tilde{\Omega}_1(\omega\mathcal{T}) = -\frac{K}{2K - \psi + j\omega}\tilde{\Omega}_1(\omega\mathcal{T}).$$

From our arguments in Section 3.8, convergence rate ψ is guaranteed if the eigenloci of $G_1(-\psi + j\omega)$ encircle -1 exactly once clockwise. This is achieved if

$$-1 \notin -\frac{K}{2K - \psi + j\omega}\tilde{\Omega}_1(\omega\mathcal{T}) \tag{C.1}$$

for all $\omega \in \mathbb{R} \setminus \{0\}$. The encirclement of -1 occurs as ω passes from negative to positive values. If $\omega < 0$, then the right hand side of (C.1) does not contain the point -1, i.e. there are no encirclements. As $\omega \to 0$, the eigenloci converge towards the eigenvalues of $G_1(-\psi) = -\frac{K}{2K-\psi}(I + D^{-1}A_\tau(-\psi))$. We show in the next paragraph that exactly one eigenvalue of $G_1(-\psi)$ satisfies $\lambda_i(G_1(-\psi)) < -1$ whereas all other eigenvalues satisfy $\lambda_i(G_1(-\psi)) > -1$ due to (3.46). Hence, exactly one eigenlocus of G_1 encircles -1 once clockwise because $\mathrm{Im}(-\frac{K}{2K-\psi-j\omega})$ changes from negative to positive values as ω changes from negative to positive values. All other eigenloci of G_1 do not encircle -1. For $\omega > 0$, condition (C.1) holds again and there are no further encirclements possible. We conclude that there is exactly one root of Δ with real part larger than $-\psi$: the zero root of Δ corresponding to the consensus solutions. Hence, consensus is reached with convergence rate ψ.

Now, we determine the eigenvalues of $G_1(-\psi) = -\frac{K}{2K-\psi}(I + D^{-1}A_\tau(-\psi))$. Consider first the eigenvalues of $D^{-1}A_\tau(-\psi)$. Remember that $A_\tau(-\psi) = [a_{ij}e^{\tau_{ij}\psi}]$ and that all delays are symmetric, i.e. $\tau_{ij} = \tau_{ji}$. Therefore, $A_\tau(-\psi)$ is real symmetric and therefore $\sigma(D^{-1}A_\tau(-\psi)) = \sigma(D^{-\frac{1}{2}}A_\tau(-\psi)D^{-\frac{1}{2}})$, where $D^{-\frac{1}{2}}A_\tau(-\psi)D^{-\frac{1}{2}}$ is real symmetric and all its eigenvalues are real. In order to bound the location of the eigenvalues of $D^{-\frac{1}{2}}A_\tau(-\psi)D^{-\frac{1}{2}}$, we use Theorem 4.3.1 in Horn and Johnson (1985) that states

$$\lambda_i(M) + \min_j \lambda_j(\tilde{M}) \leq \lambda_i(M + \tilde{M}) \leq \lambda_i(M) + \max_j \lambda_j(\tilde{M}),$$

where $M, \tilde{M}$ are real, symmetric matrices, and the eigenvalues of M and $M + \tilde{M}$ are ordered such that $\lambda_i \leq \lambda_{i+1}$. Now, we introduce the following decomposition

$$D^{-\frac{1}{2}}A_\tau(-\psi)D^{-\frac{1}{2}} = D^{-\frac{1}{2}}AD^{-\frac{1}{2}} + D^{-\frac{1}{2}}\tilde{A}(\psi)D^{-\frac{1}{2}},$$

where A is the adjacency matrix of the underlying graph and $\tilde{A}(\psi) = [a_{ij}(e^{\tau_{ij}\psi} - 1)]$. Computing the field of values of $D^{-\frac{1}{2}}\tilde{A}(\psi)D^{-\frac{1}{2}}$ in a similar way as in the proof of Lemma 3.5, we obtain $\lambda_i(D^{-\frac{1}{2}}\tilde{A}(\psi)D^{-\frac{1}{2}}) \in [1 - e^{\mathcal{T}\psi}, e^{\mathcal{T}\psi} - 1]$.

Remember that $D^{-1}A$ has a single eigenvalue $+1$ that corresponds to the zero root of Δ_0. The second largest eigenvalue of $D^{-1}A$ is $1 - \lambda_2$, where λ_2 is the second smallest eigenvalue of $\overline{L} = I - D^{-1}A$. The corresponding eigenvalue of $D^{-\frac{1}{2}}A_\tau(-\psi)D^{-\frac{1}{2}}$ satisfies $\lambda_i(D^{-\frac{1}{2}}A_\tau(-\psi)D^{-\frac{1}{2}}) \leq \lambda_i(D^{-\frac{1}{2}}AD^{-\frac{1}{2}}) + \max_j \lambda_j(D^{-\frac{1}{2}}\tilde{A}(\psi)D^{-\frac{1}{2}}) \leq 1 - \lambda_2 + e^{\mathcal{T}\psi} - 1 = e^{\mathcal{T}\psi} - \lambda_2 \leq e^{\mathcal{T}\psi} - \overline{\lambda}_2$, where $\overline{\lambda}_2 \in (0, 1)$ is a lower bound on λ_2. Thus, the corresponding eigenvalue of $G_1(-\psi)$ satisfies $\lambda_2(G_1(-\psi)) \geq -\frac{K}{2K-\psi}(1 + e^{\mathcal{T}\psi} - \overline{\lambda}_2)$, i.e. $\lambda_2(G_1(-\psi)) > -1$ due to (3.46). Thus, $n-1$ eigenloci of $G_1(-\psi)$ do not encircle -1. We know however that $\tilde{\Delta}$ has at least one right half plane pole, i.e. there must be one eigenvalue of $G_1(-\psi)$ that encircles -1, which is the one corresponding to the eigenvalue $+1$ of $D^{-1}A$. Note that this eigenvalue satisfies $\lambda_1(G_1(-\psi)) \in [-\frac{K}{2K-\psi}(1 + e^{\mathcal{T}\psi}), -\frac{K}{2K-\psi}(3 - e^{\mathcal{T}\psi})]$, which includes eigenvalues smaller than -1.

Finally, we prove that (C.1) can be replaced by (3.46). First, rewrite (C.1) in the following way: $\frac{2K-\psi+j\omega}{K} \notin \tilde{\Omega}_1(\omega\mathcal{T})$. Note that the real part of $2 + H^{-1}(-\psi + j\omega)$ is constant and we may reduce our analysis to the imaginary part, see Figure C.3(a) for an illustration. Condition (C.1) holds if $\frac{\omega}{K}$ is greater than the imaginary part of all $z \in \tilde{\Omega}_1(\omega\mathcal{T})$ with $\mathrm{Re}\{z\} = \frac{2K-\psi}{K}$, see Figure C.3(a). We have

$$\frac{\omega}{K} > \left(1 + e^{\psi\mathcal{T}} - \frac{2K-\psi}{K}\right)\tan\frac{\omega\mathcal{T}}{2} = \left(e^{\psi\mathcal{T}} - \frac{K-\psi}{K}\right)\tan\frac{\omega\mathcal{T}}{2} \tag{C.2}$$

as long as $\omega\mathcal{T} < \vartheta = \pi - \arccos\frac{K-\psi}{Ke^{\psi\mathcal{T}}}$, see Figure C.3(a). Since $K > \psi$, we have $\frac{K-\psi}{Ke^{\psi\mathcal{T}}} \in (0, 1)$ and therefore the arcus cosine is always well defined. The condition $\omega\mathcal{T} < \vartheta$ is transformed into

$$\cos(\omega\mathcal{T}) > \cos(\vartheta) = -\frac{K-\psi}{Ke^{\psi\mathcal{T}}}. \tag{C.3}$$

Hence, (C.2) holds if $\frac{\omega}{K} > e^{\psi\mathcal{T}}\sin\omega\mathcal{T}$, because

$$e^{\psi\mathcal{T}}\sin\omega\mathcal{T} = e^{\psi\mathcal{T}}\left(1 + \cos(\omega\mathcal{T})\right)\tan\frac{\omega\mathcal{T}}{2} > \left(e^{\psi\mathcal{T}} - \frac{K-\psi}{K}\right)\tan\frac{\omega\mathcal{T}}{2},$$

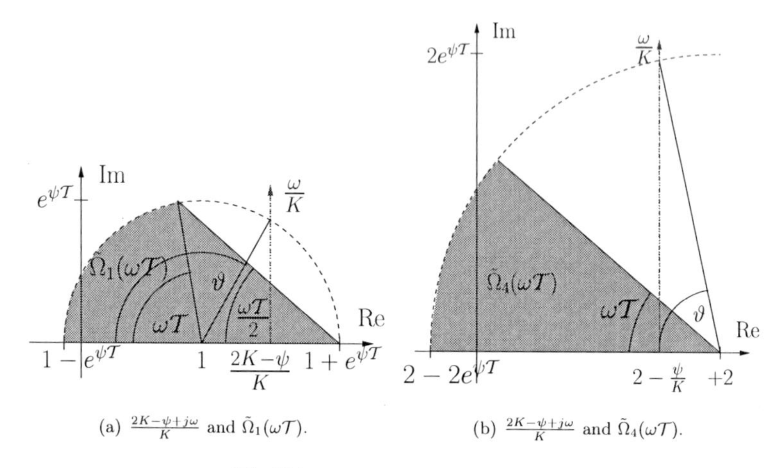

(a) $\frac{2K-\psi+j\omega}{K}$ and $\tilde{\Omega}_1(\omega\mathcal{T})$.

(b) $\frac{2K-\psi+j\omega}{K}$ and $\tilde{\Omega}_4(\omega\mathcal{T})$.

Figure C.3: Illustration of $\frac{2K-\psi+j\omega}{K}$ as red dash-dotted line and sets $\tilde{\Omega}_1$ and $\tilde{\Omega}_4$ for the proofs of Theorem 3.17 and 3.18.

due to (C.3). Hence, (C.1) holds if

$$\begin{aligned} \omega &> K e^{\psi\mathcal{T}} \sin\omega\mathcal{T} \qquad \text{or} \\ 1 &> \mathcal{T} K e^{\psi\mathcal{T}} \frac{\sin\omega\mathcal{T}}{\omega\mathcal{T}}, \end{aligned}$$

which holds if (3.46) is satisfied because $\frac{\sin\omega\mathcal{T}}{\omega\mathcal{T}} \leq 1$.

C.6 Proof of Theorem 3.18

For MAS with identical symmetric self-delay, the field of values of $2I - \Gamma_4(s)$ for $s = -\psi + j\omega$ is

$$F(2I - D^{-\frac{1}{2}} L_\tau(-\psi + j\omega) D^{-\frac{1}{2}}) \subseteq \tilde{\Omega}_4(\omega\mathcal{T}) = \mathrm{Co}\left\{2 - 2e^{\psi\mathcal{T}} e^{-j\chi}, 2 : \chi \in [0, \omega\mathcal{T}]\right\},$$

which follows from the proof of Lemma 3.5. Similar as in the proof of Theorem 3.17,

$$-1 \notin -\frac{K}{2K-\psi+j\omega}\tilde{\Omega}_4(\omega\mathcal{T}) \qquad \text{for all } \omega \in \mathbb{R}\setminus\{0\}, \tag{C.4}$$

guarantees that the eigenloci of $G_4(-\psi + j\omega)$ encircle -1 exactly once clockwise. Note however that the eigenloci converge towards the eigenvalues of $G_4(-\psi) = -\frac{K}{2K-\psi}\left(2I - D^{-}\right.$ as $\omega \to 0$.

We first consider the spectrum of $D^{-1}L_\tau(-\psi)$ as in the proof of Theorem 3.17. Since the delays are symmetric, we know that $L_\tau(-\psi)$ and $D^{-\frac{1}{2}}L_\tau(-\psi)D^{-\frac{1}{2}}$ are real, symmetric, and, moreover, $\sigma(D^{-1}L_\tau(-\psi)) = \sigma(D^{-\frac{1}{2}}L_\tau(-\psi)D^{-\frac{1}{2}})$. We propose the following decomposition

$$D^{-\frac{1}{2}}L_\tau(-\psi)D^{-\frac{1}{2}} = D^{-\frac{1}{2}}LD^{-\frac{1}{2}} + D^{-\frac{1}{2}}\tilde{L}(\psi)D^{-\frac{1}{2}},$$

where L is the Laplacian matrix of the underlying graph and $\tilde{L}(\psi) = [\tilde{l}_{ij}]$ satisfies $\tilde{l}_{ij} = a_{ij}(e^{\tau_{ij}\psi} - 1)$ for $i \neq j$ and $\tilde{l}_{ii} = \sum_{j=1}^{N} a_{ij}(e^{\tau_{ij}\psi} - 1)$. Computing the field of values of $D^{-\frac{1}{2}}\tilde{L}(\psi)D^{-\frac{1}{2}}$ in a similar way as in the proof of Lemma 3.5, we obtain $\lambda_i(D^{-\frac{1}{2}}\tilde{L}(\psi)D^{-\frac{1}{2}}) \in [0, 2e^{\mathcal{T}\psi} - 2]$, i.e. the eigenvalues of $D^{-1}L_\tau(-\psi)$ are not smaller than the eigenvalues of $D^{-1}L$. Consequently, $\psi < \lambda_2 K$ implies that exactly one eigenlocus encircles the point -1 once clockwise and all other eigenloci do not encircle -1. Note that we do not need a condition on ψ like the first part of (3.46) in this case.

Now, we prove that (C.4) holds if (3.47) holds. We transform condition (C.4) as follows: $\frac{2K-\psi+j\omega}{K} \notin \tilde{\Omega}_4(\omega\mathcal{T})$, see Figure C.3(b) for an illustration. The real part of $2 + H^{-1}(-\psi + j\omega)$ is again constant and we may reduce our analysis to the imaginary part. Condition (C.4) holds if $\frac{\omega}{K}$ is greater than the imaginary part of all $z \in \tilde{\Omega}_4(\omega\mathcal{T})$ such that $\mathrm{Re}\{z\} = \frac{2K-\psi}{K}$, see Figure C.3(b). In other words,

$$\frac{\omega}{K} > \left(2 - \frac{2K-\psi}{K}\right)\tan\omega\mathcal{T} = \frac{\psi}{K}\tan\omega\mathcal{T}, \tag{C.5}$$

for sufficiently small ω such that $\omega\mathcal{T} < \vartheta = \arccos\frac{\psi}{2Ke^{\psi\mathcal{T}}}$. Taking the cosine on both sides of $\omega\mathcal{T} < \vartheta$ yields

$$2e^{\psi\mathcal{T}}\cos\omega\mathcal{T} > \frac{\psi}{K}.$$

Hence, (C.5) holds if $\frac{\omega}{K} > 2e^{\psi\mathcal{T}}\sin\omega\mathcal{T}$ because

$$2e^{\psi\mathcal{T}}\sin\omega\mathcal{T} = 2e^{\psi\mathcal{T}}\cos\omega\mathcal{T}\tan\omega\mathcal{T} > \frac{\psi}{K}\tan\omega\mathcal{T}.$$

Thus, (C.4) holds if $1 > 2K\mathcal{T}e^{\psi\mathcal{T}}\frac{\sin\omega\mathcal{T}}{\omega\mathcal{T}}$, which holds if (C.4) is satisfied.

C.7 Proof of Theorem 3.19

For MAS with different symmetric self-delay, the field of values of $2I - \Gamma_3(s)$ for $s = -\psi + j\omega$ is composed of $F(D^{-\frac{1}{2}}D_T(-\psi + j\omega)D^{-\frac{1}{2}}) \subset \mathrm{Co}\left\{\vartheta e^{-j\varphi} : \varphi \in [0, \omega\mathcal{T}], \vartheta \in [1, e^{\psi\mathcal{T}}]\right\}$ and $F(D^{-\frac{1}{2}}A_\tau(-\psi + j\omega)D^{-\frac{1}{2}}) \subset e^{\psi\mathcal{T}}\Omega_1(\omega\mathcal{T})$, and therefore, we have

$$\begin{aligned}\sigma(2I - \Gamma_3(-\psi + j\omega)) &\subset \tilde{\Omega}_3(\omega\mathcal{T}) \\ &\subseteq \mathrm{Co}\Big\{2 - \vartheta_1 e^{-j\chi_1} + e^{\psi\mathcal{T}}e^{-j\chi_2}, 2 - \vartheta_2 e^{-j\chi_3} - e^{\psi\mathcal{T}}e^{-j\chi_4} : \\ &\qquad \chi_1, \chi_2, \chi_3, \chi_4 \in [0, \omega\mathcal{T}], \vartheta_1, \vartheta_2 \in [1, e^{\psi\mathcal{T}}]\Big\}.\end{aligned} \tag{C.6}$$

Hence, the return ratio satisfies $\sigma\left(G_3(-\psi+j\omega)\right) \subseteq -\frac{K}{2K-\psi+j\omega}\tilde{\Omega}_3(\omega\mathcal{T})$.

As in the proof of Theorem 3.17, convergence rate ψ is guaranteed if the eigenloci of $G_3(-\psi+j\omega)$ encircle -1 exactly once clockwise. This is achieved if

$$-1 \notin -\frac{K}{2K-\psi+j\omega}\tilde{\Omega}_3(\omega\mathcal{T}) \qquad \text{for all } \omega \in \mathbb{R}\setminus\{0\}. \tag{C.7}$$

As $\omega \to 0$, the eigenloci converge towards the eigenvalues of

$$G_3(-\psi) = -\frac{K}{2K-\psi}\left(2I - D^{-1}(D_T(-\psi) - A_\tau(-\psi))\right).$$

As before, we introduce the following decomposition

$$D^{-\frac{1}{2}}(D_T(-\psi) - A_\tau(-\psi))D^{-\frac{1}{2}} = D^{-\frac{1}{2}}LD^{-\frac{1}{2}} + D^{-\frac{1}{2}}\hat{L}(\psi)D^{-\frac{1}{2}},$$

where L is the Laplacian matrix and $\hat{L}(\psi) = [\hat{l}_{ij}]$ satisfies $\hat{l}_{ij} = a_{ij}(e^{T_{ij}\psi}-1)$ for $i \neq j$ and $\hat{l}_{ii} = \sum_{j=1}^{N} a_{ij}(e^{T_{ij}\psi}-1)$. Computing the field of values in a similar way as before, we obtain $\lambda_i(D^{-\frac{1}{2}}\tilde{L}_T(\psi)D^{-\frac{1}{2}}) \in [1-e^{\mathcal{T}\psi}, 2e^{\mathcal{T}\psi}-2]$.

Using similar arguments as in the proof of Theorem 3.17, we see that the second smallest eigenvalue of $D^{-\frac{1}{2}}(D_T(-\psi) - A_\tau(-\psi))D^{-\frac{1}{2}}$ is larger than $\lambda_2 + 1 - e^{\mathcal{T}\psi}$. The corresponding eigenvalue of $G_3(-\psi)$ satisfies $\lambda_2(G_3(-\psi)) \geq -\frac{K}{2K-\psi}(1+e^{\mathcal{T}\psi}-\overline{\lambda}_2)$, i.e. $\lambda_2(G_3(-\psi)) > -1$ due to (3.48). Therefore, one eigenvalue of $G_3(-\psi)$ is less than -1 and all other eigenvalues are larger than -1. Thus, exactly one eigenlocus of $G_3(-\psi+j\omega)$ encircles the point -1 once clockwise as ω changes from negative to positive values and all other eigenloci do not encircle -1.

Finally, we show that (C.7) holds if (3.48) holds. We transform (C.7) into $\frac{2K-\psi+j\omega}{K} \notin \tilde{\Omega}_3(\omega\mathcal{T})$, see Figure C.4 for an illustration. Again, we may reduce our analysis to the imaginary part. Condition (C.7) is satisfied if $\frac{\omega}{K}$ is larger than the imaginary part of all $z \in \tilde{\Omega}_3(\omega\mathcal{T})$ such that $\mathrm{Re}\{z\} = 2 - \frac{\psi}{K}$, see Figure C.4. The imaginary part of these z is bounded by

$$\begin{aligned}\mathrm{Im}(z) &\leq e^{\psi\mathcal{T}}\left(\sin(\omega\mathcal{T}) + \left(1-\cos(\omega\mathcal{T}) + \frac{\psi}{Ke^{\psi\mathcal{T}}}\right)\tan\left(\frac{\omega\mathcal{T}}{2}\right)\right)\\ &= e^{\psi\mathcal{T}}\sin(\omega\mathcal{T})\left(1 + \frac{1-\cos\omega\mathcal{T} + \frac{\psi}{Ke^{\psi\mathcal{T}}}}{1+\cos\omega\mathcal{T}}\right).\end{aligned}$$

Hence, (C.7) holds if $\frac{\omega}{K} > \mathrm{Im}(z)$ as long as $\frac{2K-\psi+j\omega}{K}$ is inside the circle containing $\tilde{\Omega}_3$. Note that $\frac{2K-\psi+j\omega}{K}$ leaves this circle at the point ρ indicated in Figure C.4. The condition $\frac{\omega}{K} > \mathrm{Im}(z)$ can be reformulated as

$$\frac{1}{K\mathcal{T}} > e^{\psi\mathcal{T}}\frac{\sin(\omega\mathcal{T})}{\omega\mathcal{T}}\left(1 + \frac{1-\cos\omega\mathcal{T} + \frac{\psi}{Ke^{\psi\mathcal{T}}}}{1+\cos\omega\mathcal{T}}\right) = f(\omega\mathcal{T}).$$

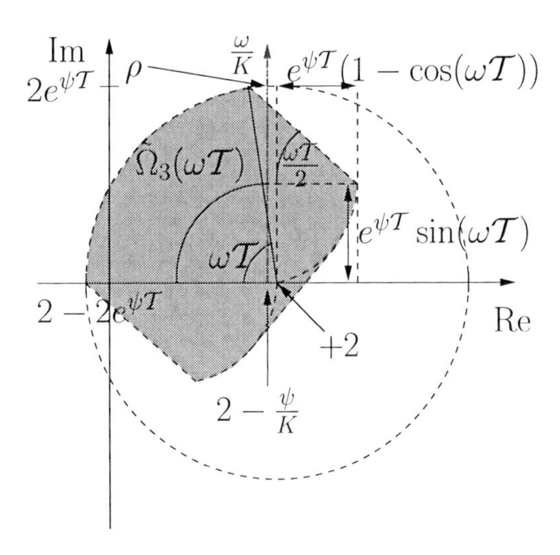

Figure C.4: Illustration of $\frac{2K-\psi+j\omega}{K}$ as red dash-dotted line and set $\tilde{\Omega}_3$ for the proof of Theorem 3.19.

The derivative of f with respect to $\omega\mathcal{T}$ is

$$\frac{df(\omega\mathcal{T})}{d\omega\mathcal{T}} = e^{\psi\mathcal{T}}\left(2+\frac{\psi}{Ke^{\psi\mathcal{T}}}\right)\frac{\omega\mathcal{T}-\sin(\omega\mathcal{T})}{(\omega\mathcal{T})^2(1+\cos(\omega\mathcal{T}))},$$

i.e. f is strictly increasing for $\omega\mathcal{T}\in(0,\frac{\pi}{2})$. Hence, if $\frac{2K-\psi+j\omega}{K}$ reaches ρ before $\tilde{\Omega}_3$, i.e. before $\rho\in\tilde{\Omega}_3$, then we know that $\frac{\omega}{K}>\mathrm{Im}(z)$ for all ω. Thus, we only have to check one ω. Denote $\omega^*>0$ the smallest frequency where $\rho\in\tilde{\Omega}_3(\omega^*)$. Hence, $\frac{\omega^*}{K}>2e^{\psi\mathcal{T}}\sin(\omega^*\mathcal{T})$ implies (C.7), see Figure C.4. Since

$$1>2\mathcal{T}Ke^{\psi\mathcal{T}}\geq 2\mathcal{T}Ke^{\psi\mathcal{T}}\frac{\sin(\omega^*\mathcal{T})}{\omega^*\mathcal{T}},$$

we conclude that (C.7) holds if (3.48) holds.

Bibliography

J. Almeida, C. Silvestre, A. Pascoal, and P. J. Antsaklis. Continuous-time consensus with discrete-time communication. In *Proc. Europ. Contr. Conf.*, pages 749–754, Budapest, Hungary, 2009.

P. Antsaklis and J. Baillieul. Special issue on technology of networked control systems, *Proc. of the IEEE*, 95(1):5–312, 2007.

M. Arcak. Passivity as a design tool for group coordination. *IEEE Trans. Autom. Control*, 52(8):1380–1390, 2007.

F. M. Atay. Synchronization and amplitude death in coupled limit cycle oscillators with time delays. In J. J. Loiseau, W. Michiels, S.-I. Niculescu, and R. Sipahi, editors, *Topics in Time Delay Systems: Analysis, Algorithms, and Control*, volume 388 of *LNCIS*, pages 383–389. Springer, Berlin, Germany, 2009.

F. M. Atay and Ö. Karabacak. Stability of coupled map networks with delays. *SIAM J. Applied Dynamical Systems*, 5(3):508–527, 2006.

M. Bando, K. Hasebe, K. Nakanishi, A. Shibata, and Y. Sugiyama. Structure stability of congestion in traffic dynamics. *Japan Journal of Industrial and Applied Mathematics*, 11(2):203–2245, 1994.

M. Bando, K. Hasebe, K. Nakanishi, and A. Nakayama. Analysis of optimal velocity model with explicit delay. *Physical Review E*, 58(5):5429–5435, 1998.

J. Bischoff. Kybernetik: Roboter lernen Teamarbeit. *Geo Magazin Deutschland*, 7: 46–60, 2009.

P.-A. Bliman and G. Ferrari-Trecate. Average consensus problems in networks of agents with delayed communications. *Automatica*, 44(8):1985–1995, 2008.

A.B. Bondi. Characteristics of scalability and their impact on performance. In *Proc. 2nd intern. Workshop on Software and Performance*, pages 195–203, Ontario, Canada, 2000.

S. Boyd and L. Vandenberghe. *Convex Optimization.* Cambridge University Press, New York, USA, 2004.

M. Brackstone and M. McDonald. Car-following: A historical review. *Transporation Research Part F*, 2(4):181–196, 1999.

D. Bratsun, D. Volfson, L.S. Tsimring, and J. Hasty. Delay-induced stochastic oscillations in gene regulation. *Proc. National Academy of Science U.S.A.*, 102(41): 14593–14598, 2005.

D. Breda, R. Vermiglio, and S. Maset. Tool for robust analysis and characteristic equation of delay differential equations (trace-dde), 2010. Available from http://users.dimi.uniud.it/ dimitri.breda/software.html .

F. Bullo, J. Cortés, and S. Martínez. *Distributed Control of Robotic Networks*. Princeton University Press, 2009.

M. Cao, A. S. Morse, and B. D. O. Anderson. Reaching a consensus in a dynamically changing environment: A graphical approach & convergence ranges, measurement delays, and asynchronous events. *SIAM J. Control Optim.*, 47(2):575–623, 2008.

R. Carli and F. Bullo. Quantized coordination algorithms for rendezvous and deployment. *SIAM J. Control Optim.*, 48(3):1251–1274, 2009.

D. Castro, R. Meir, and I. Yavneh. Delays and oscillations in networks of spiking neurons: a two-timescale analysis. *Neural Computation*, 21(4):1100–1124, 1999.

T. Charalambous and Y. Ariba. On the stability of power control algorithm for wireless networks in the presence of time-varying delays. In *Proc. Europ. Contr. Conf.*, pages 2936–2941, Budapest, Hungary, 2009.

T. Charalambous, I. Lestas, and G. Vinnicombe. On the stability of the Foschini-Miljanic algorithm with time-delays. In *Proc. IEEE Conf. Decision Contr.*, pages 2991–2996, Cancun, Mexico, 2008.

V. S. Chellaboina, W. M. Haddad, Q. Hui, and J. Ramakrishnan. On system state equipartitioning and semistability in network dynamical systems with arbitrary time-delays. In *Proc. IEEE Conf. Decision Contr.*, pages 3461–3466, San Diego, USA, 2006.

L. Chen and K. Aihara. A model of periodic oscillation for genetic regulatory systems. *IEEE Trans. Circuits and Systems—I: Fundamental Theory and Applications*, 49(10): 1429–1436, 2002.

N. Chopra and M. Spong. Passivity-based control of multi-agent systems. In S. Kaxamura and M. Svinin, editors, *Advances in Robot Control*, pages 107–134. Springer, Berlin, Germany, 2006.

N. Chopra and M. Spong. Output synchronization of nonlinear systems with relative degree one. In V. D. Blondel, S. P. Boyd, and H. Kimura, editors, *Recent Advances in Learning and Control*, pages 51–64. Springer, Berlin, Germany, 2008.

N. Chopra, D. M. Stipanović, and M. W. Spong. On synchronization and collision avoidance for mechanical systems. In *Proc. Amer. Contr. Conf.*, pages 3713–3718, Seattle, USA, 2008.

F. R. K. Chung. *Spectral Graph Theory.* American Mathematical Society, Providence, USA, 1997.

I. D. Couzin, J. Krause, N. R. Franks, and S. A. Levin. Effective leadership and decision-making in animal groups on the move. *Nature*, 433:513–516, 2005.

J. M. Cushing. *Integrodifferential Equations and Delay Models in Population Dynamics.* Springer, Heidelberg, 1977.

M. H. DeGroot. Reaching a consensus. *J. American Statistical Association*, 69(345): 118–121, 1974.

C. A. Desoer and Y.-T. Wang. On the generalized Nyquist stability criterion. *IEEE Trans. Autom. Control*, 25(2):187–196, 1980.

W. Dong and J. A. Farrell. Cooperative control of multiple nonholonomic mobile agents. *IEEE Trans. Autom. Control*, 53(6):1434–1448, 2008.

W. Dong and J. A. Farrell. Decentralized cooperative control of multiple nonholonomic dynamic systems with uncertainty. *Automatica*, 45(3):706–710, 2009.

M. G. Earl and S. H. Strogatz. Synchronization in oscillator networks with delayed coupling: A stability criterion. *Physical Review E*, 67(3):036204, 2003.

E. Eisenberg and D. Gale. Consensus of subjective probabilities: The pari-mutuel method. *The Annals of Mathematical Statistics*, 30(1):165–168, 1959.

X. Fan and M. Arcak. Delay robustness of a class of nonlinear systems and applications to communication networks. *IEEE Trans. Autom. Control*, 51(1):139–144, 2006.

L. Fang and P. J. Antsaklis. Information consensus of asynchronous discrete-time multi-agent systems. In *Proc. Amer. Contr. Conf.*, pages 1883–1888, Portland, USA, 2005.

J. A. Fax and R. M. Murray. Information flow and cooperative control of vehicle formations. *IEEE Trans. Autom. Control*, 49(9):1465–1476, 2004.

E. Fridman, A. Seuret, and J.-P. Richard. Robust sampled-data stabilization of linear systems: An input delay approach. *Automatica*, 40(8):1441–1446, 2004.

R. Ghabcheloo, A. P. Aguiar, A. Pascoal, and C. Silvestre. Synchronization in multi-agent systems with switching topologies and non-homogeneous communication delays. In *Proc. IEEE Conf. Decision Contr.*, pages 2327–2332, New Orleans, USA, 2007.

C. Godsil and G. Royle. *Algebraic Graph Theory.* Springer, New York, USA, 2000.

G. Goebel, U. Münz, and F. Allgöwer. Stabilization of linear systems with distributed input delay. In *Proc. Amer. Contr. Conf.*, pages 5800–5806, Baltimore, USA, 2010.

F. Gouaisbaut and D. Peaucelle. Stability of time-delay systems with non-small delay. In *Proc. IEEE Conf. Decision Contr.*, pages 840–845, San Diego, USA, 2006.

K. Gu. Refined discretized Lyapunov functional method for systems with multiple delays. *Int. J. Robust and Nonlinear Control*, 13(11):1017–1033, 2003.

K. Gu, V. L. Kharitonov, and J. Chen. *Stability of Time-Delay Systems.* Birkhäuser, Boston, 2003.

T. Haag, U. Münz, and F. Allgöwer. Comparison of different stability conditions for linear time-delay systems with incommensurate delays. In *Proc. IFAC Workshop on Time-Delay Systems*, Sinaia, Romania, 2009.

J. Hale and S. M. V. Lunel. *Introduction to Functional Differential Equations.* Springer, New York, USA, 1993.

G. Hardy, J. E. Littlewood, and G. Pólya. *Inequalities.* Cambridge Universtiy Press, Cambridge, UK, 2nd edition, 1952.

Y. Hatano and M. Mesbahi. Agreement over random networks. *IEEE Trans. Autom. Control*, 50:1867–1872, 2005.

D. Helbing. Traffic and related self-driven many-particle systems. *Reviews of Modern Physics*, 73(4):1067–1141, 2001.

R. A. Horn and C. R. Johnson. *Matrix Analysis.* Cambridge University Press, New York, USA, 1st edition, 1985.

R. A. Horn and C. R. Johnson. *Topics in Matrix Analysis.* Cambridge University Press, New York, USA, 1st edition, 1991.

X. Hu, Y. Hong, and L. Bushnell. Special issue on collective behavior and control of Multi-Agent Systems, *Asian J. Contr.*, 10(2):129–266, 2008.

C. Huygens. *Horologium Oscillatorium.* Apud F. Muguet, Parisiis, France, 1673.

A. Jadbabaie, J. Lin, and S. Morse. Coordination of groups of mobile autonomous agents using nearest neighbor rules. *IEEE Trans. Autom. Control*, 48(6):988–1001, 2003.

E. Jarlebring. *The Sprectrum of Delay-Differential Equations: Numerical Methods, Stability, and Perturbations.* PhD thesis, Techical University Carolo-Wilhelmina of Braunschweig, Germany, 2008.

U. T. Jönsson and C.-Y. Kao. A scalable robust stability criterion for systems with heterogeneous LTI components. In *Proc. Amer. Contr. Conf.*, pages 2898–2903, St. Louis, USA, 2009.

C.-Y. Kao, U. Jönsson, and H. Fujioka. Characterization of robust stability of a class of interconnected systems. *Automatica*, 45(1):217–224, 2009.

A. Kashyap, T. Başar, and R. Srikant. Quantized consensus. *Automatica*, 43(7):1192–1203, 2007.

H. K. Khalil. *Nonlinear Systems*. Prentice Hall, Upper Saddle River, NJ, 3rd edition, 2002.

Y. Kuramoto. *Chemical Oscillations, Waves, and Turbulence*. Springer, Berlin, Germany, 1984.

D. Lee and M. W. Spong. Agreement with non-uniform information delays. In *Proc. Amer. Contr. Conf.*, pages 756–761, Minneapolis, USA, 2006.

I. C. Lestas and G. Vinnicombe. Scalable robustness for consensus protocols with heterogeneous dynamics. In *Proc. IFAC World Congress*, Prague, Czech Republic, 2005.

I. C. Lestas and G. Vinnicombe. Scalable decentralized robust stability certificates for networks of interconnected heterogeneous dynamical systems. *IEEE Trans. Autom. Control*, 51(10):1613–1625, 2006.

I. C. Lestas and G. Vinnicombe. Scalable robust stability for nonsymmetric heterogeneous networks. *Automatica*, 43(4):714–723, 2007a.

I. C. Lestas and G. Vinnicombe. The S-hull approach to consensus. In *Proc. IEEE Conf. Decision Contr.*, pages 182–187, New Orleans, USA, 2007b.

P. Lin and Y. Jia. Consensus of second-order discrete-time multi-agent systems with nonuniform time-delay and dynamically changing topologies. *Automatica*, 45(9):2154–2158, 2009.

Z. Lin. *Coupled Dynamic Systems: From Structure towards Stability and Stabilizability*. PhD thesis, University of Toronto, Canada, 2006.

Z. Lin, B. Francis, and M. Maggiore. State agreement for continuous-time coupled nonlinear systems. *SIAM J. Control Optim.*, 46(1):288–307, 2007.

C.-L. Liu and Y.-P. Tian. Consensus of Multi-Agent System with diverse communication delays. In *Proc. Chin. Contr. Conf.*, pages 726–730, Zhangjiajie, Hunan, China, 2007.

C.-L. Liu and Y.-P. Tian. Coordination of multi-agent systems with communication delays. In *Proc. IFAC World Congress*, pages 10782–10787, Seoul, South Korea, 2008.

Y. Liu and K. M. Passino. Cohesive behaviors of multi-agent systems with information flow contraints. *IEEE Trans. Autom. Control*, 51(11):1734–1748, 2006.

J. J. Loiseau, W. Michiels, S.-I. Niculescu, and R. Sipahi, editors. *Topics in Time Delay Systems: Analysis, Algorithms, and Control*, volume 388 of *LNCIS*. Springer, Berlin, Germany, 2009.

I. Lopez, J. L. Piovesan, C. T. Abdallah, D. Lee, O. Martinez, M. W. Spong, and R. Sandoval. Practical issues in networked control systems. In *Proc. Amer. Contr. Conf.*, pages 4201–4206, Minneapolis, USA, 2006.

R. Lupas Scheiterer, C. Na, D. Obradovic, and G. Steindl. Synchronization performance of the precision time protocol in industrial automation networks. *IEEE Trans. Instrumentation and Measurement*, 58(6):1849–1857, 2009.

R. Mahboobi Esfanjani, M. Reble, U. Münz, S. K. Nikravesh, and F. Allgöwer. Model predictive control of constrained nonlinear time-delay systems. In *Proc. IEEE Conf. Decision Contr.*, pages 1324–1329, Shanghai, China, 2009.

C. Maier, T. Haag, U. Münz, and F. Allgöwer. Construction of quadratic Lyapunov-Krasovskii functionals for linear time-delay systems with multiple uncertain delays. In S. Sivasundaram, editor, *Mathematical Analysis and Applications in Engineering Aerospace and Sciences*, volume 5 of *Mathematical Problems in Engineering Aerospace and Sciences*. Cambridge Scientific Publishers, Cambridge, UK, 2010. (in press).

The MathWorks. SimEvents 1.2 toolbox, 2006. Available from http://www.mathworks.com/products/simevents/.

D. Mehdi, E. K. Boukas, and Z. K. Liu. Dynamical systems with multiple time-varying delays: Stability and stabilizability. *J. of Optimization Theory and Applications*, 113 (3):537–565, 2002.

M. Michiels and S.-I. Niculescu. *Stability and Stabilization of Time-Delay Systems: An Eigenvalue-Based Approach.* SIAM, Philadelphia, USA, 2007.

W. Michiels and H. Nijmeijer. Synchronization of delay-coupled nonlinear oscillators: An approach based on the stability analysis of synchronized equilibria. *Chaos*, 19(3): 033110, 2009.

W. Michiels, V. van Assche, and S.-I. Niculescu. Stabilization of time-delay systems with a controlled time-varying delay and applications. *IEEE Trans. Autom. Control*, 50(4):493–504, 2005.

W. Michiels, C.-I. Morărescu, and S.-I. Niculescu. Consensus problems with distributed delays, with application to traffic flow models. *SIAM J. Control Optim.*, 48(1):77–101, 2009.

P. Miller. Swarm theory. *National Geographic*, 7:126–147, 2007a.

P. Miller. Schwarmintelligenz. *National Geographic Deutschland*, 8:42–63, 2007b.

C.-I. Morărescu, S.-I. Niculescu, and K. Gu. Stability crossing curves of shifted gamma-distributed delay systems. *SIAM J. Applied Dynamical Systems*, 6(2):475–493, 2007.

L. Moreau. Stability of continuous-time distributed consensus algorithms. In *Proc. IEEE Conf. Decision Contr.*, pages 3998–4003, Atlantis, USA, 2004.

L. Moreau. Stability of multiagent systems with time-dependent communication links. *IEEE Trans. Autom. Control*, 50(2):169–182, 2005.

S. Mossaheb. A Nyquist type stability criterion for linear multivariable delayed systems. *Int. J. Control*, 32(5):821–847, 1980.

J. R. Moyne and D. M. Tilbury. The emergence of industrial control networks for manufacturing control, diagnostics, and safety data. *Proc. IEEE*, 95(1):29–47, 2007.

U. Münz and F. Allgöwer. $\mathcal{L}_2$-gain based controller design for linear systems with distributed delays and rational delay kernels. In *Proc. IFAC Workshop on Time-Delay Systems*, Nantes, France, 2007.

U. Münz, C. Ebenbauer, and F. Allgöwer. Stability of networked systems with multiple delays using linear programming. In *Proc. Amer. Contr. Conf.*, pages 5515–5520, New York, USA, 2007a.

U. Münz, A. Papachristodoulou, and F. Allgöwer. Multi-agent system consensus in packet-switched networks. In *Proc. Europ. Contr. Conf.*, pages 4598–4603, Kos, Greece, 2007b.

U. Münz, A. Papachristodoulou, and F. Allgöwer. Nonlinear multi-agent system consensus with time-varying delays. In *Proc. IFAC World Congress*, pages 1522–1527, Seoul, South Korea, 2008a.

U. Münz, A. Papachristodoulou, and F. Allgöwer. Delay-dependent rendezvous and flocking of large scale multi-agent systems with communication delays. In *Proc. IEEE Conf. Decision Contr.*, pages 2038–2043, Cancun, Mexico, 2008b.

U. Münz, J. M. Rieber, and F. Allgöwer. Robust stability of distributed delay systems. In *Proc. IFAC World Congress*, pages 12354–12358, Seoul, South Korea, 2008c.

U. Münz, C. Ebenbauer, T. Haag, and F. Allgöwer. Stability analysis of time-delay systems in the frequency domain using positive polynomials. *IEEE Trans. Autom. Control*, 54(5):1019–1024, 2009a.

U. Münz, A. Papachristodoulou, and F. Allgöwer. Consensus reaching in multi-agent packet-switched networks. *Int. J. Control*, 82(5):953–969, 2009b.

U. Münz, A. Papachristodoulou, and F. Allgöwer. Generalized Nyquist consensus condition for large linear multi-agent systems with communication delays. In *Proc. IEEE Conf. Decision Contr.*, pages 4765–4771, Shanghai, China, 2009c.

U. Münz, A. Papachristodoulou, and F. Allgöwer. Output consensus controller design for nonlinear relative degree one multi-agent systems with delays. In *Proc. IFAC Workshop on Time-Delay Systems*, Sinaia, Romania, 2009d.

U. Münz, A. Papachristodoulou, and F. Allgöwer. Generalized Nyquist consensus condition for linear multi-agent systems with heterogeneous delays. In *Proc. IFAC Workshop on Estimation and Control of Networked Systems*, pages 24–29, Venice, Italy, 2009e.

U. Münz, J. M. Rieber, and F. Allgöwer. Robust stabilization and H-infinity control of uncertain distributed delay systems. In J. J. Loiseau, W. Michiels, S.-I. Niculescu, and R. Sipahi, editors, *Topics in Time Delay Systems: Analysis, Algorithms, and Control*, volume 388 of *LNCIS*, pages 221–231. Springer, Berlin, Germany, 2009f.

U. Münz, A. Papachristodoulou, and F. Allgöwer. Consensus controller design for nonlinear relative degree two multi-agent systems with communication constraints, 2010a. IEEE Trans. Autom. Contr. (in press).

U. Münz, A. Papachristodoulou, and F. Allgöwer. Robust rendezvous of heterogeneous Lagrange systems on packet-switched networks. *at – Automatisierungstechnik*, 58(4): 184–191, 2010b.

U. Münz, A. Papachristodoulou, and F. Allgöwer. Delay robustness in consensus problems. *Automatica*, 46(8):1252–1265, 2010c.

R. M. Murray, editor. *Control in an Information Rich World: Report of the Panel on Future Directions in Control, Dynamics, and Systems*. SIAM, 2002.

A. Nedić and A. Ozdaglar. Convergence rate for consensus with delays. *J. Global Optim.*, 47(3):437–456, 2010.

S.-I. Niculescu. *Delay Effects on Stability: A Robust Control Approach*. Springer, London, UK, 2001.

B. Novák and J.J. Tyson. Design principles of biochemical oscillators. *Nature Reviews Molecular Cell Biology*, 9(12):981–991, 2008.

R. Olfati-Saber. Ultrafast consensus in small-world networks. In *Proc. Amer. Contr. Conf.*, pages 2371–2378, Portland, USA, 2005.

R. Olfati-Saber. Flocking for multi-agent dynamic systems: Algorithms and theory. *IEEE Trans. Autom. Control*, 51(3):401–420, 2006.

R. Olfati-Saber and R. M. Murray. Consensus problem in networks of agents with switching topology and time-delays. *IEEE Trans. Autom. Control*, 49(9):1520–1533, 2004.

R. Olfati-Saber and J. S. Shamma. Consensus filters for sensor networks and distributed sensor fusion. In *Proc. IEEE Conf. Decision Contr. and Europ. Contr. Conf.*, pages 6698–6703, Seville, Spain, 2005.

R. Olfati-Saber, J. A. Fax, and R. M. Murray. Consensus and cooperation in networked multi-agent systems. *Proc. IEEE*, 95(1):215–233, 2007.

G. Orosz. *The Dynamics of Vehicular Trafic with Drivers' Reaction Time Delay*. PhD thesis, University of Bristol, UK, 2005.

R. Ortega, A. Loría, P. J. Nicklasson, and H. Sira-Ramírez. *Passivity-Based Control of Euler-Lagrange Systems*. Springer, London, UK, 1998.

A. Papachristodoulou and A. Jadbabaie. Synchronization in oscillator networks: Switching topologies and non-homogeneous delays. In *Proc. IEEE Conf. Decision Contr. and Europ. Contr. Conf.*, pages 5692–5697, Seville, Spain, 2005.

A. Papachristodoulou and A. Jadbabaie. Synchronization of oscillator networks with heterogeneous delays, switching topologies and nonlinear dynamics. In *Proc. IEEE Conf. Decision Contr.*, pages 4307–4312, San Diego, USA, 2006.

A. Papachristodoulou and A. Jadbabaie. Delay robustness of nonlinear internet congestion control schemes. *IEEE Trans. Autom. Control*, 55(6):1421–1427, 2010.

A. Papachristodoulou and M. M. Peet. Global stability analysis of primal internet congestion control schemes with heterogeneous delays. In *Proc. IFAC World Congress*, pages 2913–2918, Seoul, South Korea, 2008.

A. Papachristodoulou, A. Jadbabaie, and U. Münz. Effects of delay in multi-agent consensus and oscillatior synchronization. *IEEE Trans. Autom. Control*, 55(6):1471–1477, 2010.

M. Pavella and P. G. Murthy, editors. *Transient Stability of Power Systems: Theory and Practice*. John Wiley & Sons, West Sussex, UK, 1994.

C. Pöppe. Führerpersönlichkeit und Herdentrieb. *Spektrum der Wissenschaft*, 9:22–24, 2005.

Z. Qu. *Cooperative Control of Dynamical Systems*. Springer, London, UK, 2009.

N. Radde. The impact of time delays on the robustness of biological oscillators and the effect of bifurcations on the inverse problem. *EURASIP J. Bioinformatics and Systems Biology*, 2009:327503, 2009.

W. Ren and R. W. Beard. *Distributed Consensus in Multi-Vehicle Cooperative Control*. Springer, London, UK, 2008.

C. W. Reynolds. Flocks, herds, and schools: A distributed behavioral model. *Computer Graphics*, 21(4):25–34, 1987.

J.-P. Richard. Time-delay systems: An overview of some recent advances and open problems. *Automatica*, 39(10):1667–1694, 2003.

O. Roesch, H. Roth, and S.-I. Niculescu. Remote control of mechatronic systems over communication networks. In *Proc. Int. Conf. Mechatronics and Automation*, pages 1648–1653, Niagara Falls, Canada, 2005.

S. Roy, A. Saberi, and A. Stoorvogel. Special issue on communicating-agent networks, *International Journal of Robust and Nonlinear Control*, vol. 17, no. 10–11, pp. 897–1066, 2007.

S. Salza, M. Draoli, C. Gaibisso, A. L. Palma, and R. Puccinelli. Methods and tools for the objective evaluation of voice-over-IP communications. In *Proceedings of the 10th Annual Internet Society Conference*, Yokohama, Japan, 2000. Available from http://www.isoc.org/inet2000/cdproceedings/1i/1i_2.htm.

A. Sarlette. *Geometry and Symmetries in Coordination Control.* PhD thesis, University of Liège, Belgium, 2009.

T. Scheper, D. Klinkenberg, C. Pennartz, and J. van Pelt. A mathematical model for the intracellular circadian rhythm generator. *J. Neuroscience*, 19(1):40–47, 1999.

G. S. Schmidt, U. Münz, and F. Allgöwer. Multi-agent speed consensus via delayed position feedback with application to kuramoto oscillators. In *Proc. Europ. Contr. Conf.*, pages 2464–2469, Budapest, Hungary, 2009.

G. S. Schmidt, J. Wu, U. Münz, and F. Allgöwer. Consensus in bistable and multistable multi-agent systems, 2010. IEEE Conf. Decision Contr., Atlanta, USA (accepted).

G. Scutari, S. Barbarossa, and L. Pescosolido. Distributed decision through self-synchronizing sensor networks in the presence of propagation delays and asymmetric channels. *IEEE Trans. Signal Proc.*, 56(4):1667–1684, 2008.

R. Sepulchre, M Janković, and P. Kokotović. *Constructive Nonlinear Control.* Springer, London, 1997.

R. Sepulchre, D. Paley, and N. E. Leonard. Stabilization of planar collective motion with limited communication. *IEEE Trans. Autom. Control*, 53(3):706–719, 2008.

A. Seuret, D. V. Dimarogonas, and K. H. Johansson. Consensus under communication delays. In *Proc. IEEE Conf. Decision Contr.*, pages 4922–4927, Cancun, Mexico, 2008.

J. S. Shamma, editor. *Cooperative Control of Distributed Multi-Agent Systems.* John Wiley & Sons, West Sussex, UK, 2007.

T. Shima and P. R. Pagilla. Special issue on analysis and control of multi-agent dynamic systems, *J. Dynamic Systems, Measurement, and Control*, 129(5):569–754, 2007.

T. Shima and S. Rasmussen, editors. *UAV Cooperative Decision and Control: Challenges and Practical Approaches.* SIAM, Philadelphia, USA, 2009.

R. Sipahi and S.-I. Niculescu. A survey of deterministic time delay traffic flow models. In *Proc. IFAC Workshop on Time-Delay Systems*, Nantes, France, 2007.

S. Skogestad and I. Postlethwaite. *Multivariable Feedback Control Analysis and Design.* Wiley, New York, 2004.

R. Srikant. *The Mathematics of Internet Congestion Control.* Birkhäuser, Boston, USA, 2004.

N. Stefanovic and L. Pavel. An analysis of stability with time-delay of link level power control in optical networks. *Automatica*, 45(1):149–154, 2009a.

N. Stefanovic and L. Pavel. A stability analysis with time-delay of primal-dual power control in optical links. *Automatica*, 45(5):1319–1325, 2009b.

S. H. Strogatz. Nonlinear dynamics: Death by delay. *Nature*, 394:316–317, 1998.

S. H. Strogatz. From Kuramoto to Crawford: Exploring the onset of synchronization in populations of coupled oscillators. *Physica D*, 143(1):1–20, 2000.

S. H. Strogatz. *SYNC: The Emerging Science of Spontaneous Order*. Hyperion Press, New York, 2003.

S. H. Strogatz. *Synchron: Vom rätselhaften Rhythmus der Natur*. Berlin Verlag, Berlin, 2004.

Y. Sun and M. D. Lemmon. Convergence of consensus filtering under network throughput limitations. In *Proc. IEEE Conf. Decision Contr.*, pages 2277–2282, New Orleans, USA, 2007.

A. Tahbaz-Salehi and A. Jadbabaie. A necessary and sufficient condition for consensus over random networks. *IEEE Trans. Autom. Control*, 53(3):791–795, 2008.

H. G. Tanner, A. Jadbabaie, and G. J. Pappas. Flocking in fixed and switching networks. *IEEE Trans. Autom. Control*, 52(5):863–868, 2007.

Y.-P. Tian and C.-L. Liu. Consensus of mutli-agent systems with diverse input and communication delays. *IEEE Trans. Autom. Control*, 53(9):2122–2128, 2008.

J. Tsitsiklis. *Problems in Decentralized Decision Making and Computation*. PhD thesis, Massachusetts Institute of Technology, 1984.

J. J. Tyson. Biochemical oscillations. In C. P. Fall, E. S. Marland, J. M. Wagner, and J.J. Tyson, editors, *Computational Cell Biology*, volume 20 of *Interdisciplinary Applied Mathematics*, pages 230–260. Springer series, Berlin, Germany, 2005.

T. Vámos. Cooperative systems based on non-cooperative people. *IEEE Control Systems Magazine*, 3(3):9–14, 1983.

Verein Deutscher Ingenieure (VDI). Agentensysteme in der Automatisierungstechnik — Grundlagen, VDI-Richtlinie VDI/VDE 2653 Blatt 1, 2009.

J. Wang and N. Elia. Consensus over networks with dynamic channels. In *Proc. Amer. Contr. Conf.*, pages 2637–2642, Seattle, USA, 2008.

R. Wang, T. Zhou, Z. Jing, and L. Chen. Modelling periodic oscillation of biological systems with multiple timescale networks. *IEE Proc. Systems Biology*, 1(1):71–84, 2004.

R. Wang, L. Chen, and K. Aihara. Construction of genetic oscillators with interlocked feedback networks. *Journal of Theoretical Biology*, 242(2):454–463, 2006.

R. Wang, L. Chen, and K. Aihara. Detection of cellular rhythms and global stability within interlocked feedback systems. *Mathematical Bioscience*, 209:171–189, 2007.

P. Wieland, J.-S. Kim, H. Scheu, and F. Allgöwer. On consensus in multi-agent systems with linear high-order agents. In *Proc. IFAC World Congress*, pages 1541–1546, Seoul, South Korea, 2008.

P. Wieland, J.-S. Kim, and F. Allgöwer. On topology and dynamics of consensus among linear high-order agents. *Int. J. Systems Science*, 2010. (in press).

J. Wu. Consensus in multistable multi-agent systems. Master's thesis, Institute for Systems Theory and Automatic Control, University of Stuttgart, Germany, 2009.

F. Xiao and L. Wang. State consensus for multi-agent systems with switching topologies and time-varying delays. *Int. J. Control*, 79(10):1277–1284, 2006.

F. Xiao and L. Wang. Consensus protocols for discrete-time multi-agent systems with time-varying delays. *Automatica*, 44(10):2577–2582, 2008a.

F. Xiao and L. Wang. Asynchronous consensus in continuous-time multi-agent systems with switching topology and time-varying delays. *IEEE Trans. Autom. Control*, 53 (8):1804–1816, 2008b.

W. Yang, A. Bertozzi, and X. Wang. Stability of a second order consensus algorithm with time delay. In *Proc. IEEE Conf. Decision Contr.*, pages 2962–2931, Cancun, Mexico, 2008.

M. K. S. Yeung and S. H. Strogatz. Time delay in the Kuramoto model of coupled oscillators. *Physical Review Letters*, 82(3):648–651, 1999.

Index